Chatham House Report for the Mov

Glada Lahn and Owen Grafham
Foreword by Kofi Annan
November 2015

# Heat, Light and Power for Refugees

## Saving Lives, Reducing Costs

CHATHAM HOUSE
The Royal Institute of International Affairs

**Chatham House,** the Royal Institute of International Affairs, is an independent policy institute based in London. Our mission is to help build a sustainably secure, prosperous and just world.

The Royal Institute of International Affairs

Chatham House
10 St James's Square
London SW1Y 4LE
T: +44 (0) 20 7957 5700
F: + 44 (0) 20 7957 5710
**www.chathamhouse.org**

Charity Registration No. 208223



ISBN 978 1 78413 095 4

A catalogue record for this title is available from the British Library.

Cover image: A migrant girl looks at a light illuminating the camp site of refugees and migrants who spend the night on the street after their arrival at the Greek island of Lesbos after crossing the Aegean Sea from Turkey on 4 October 2015. Europe is grappling with its biggest migration challenge since the Second World War, with the main surge coming from civil war-torn Syria.
Ccopyright © Aris Messinis/AFP/Getty Images

Typeset by Soapbox, www.soapbox.co.uk

Printed and bound in Great Britain by City Print (Milton Keynes) Ltd

This publication is printed on recycled paper

# Contents

# About the Authors

**Glada Lahn** is a senior research fellow with the Energy, Environment and Resources Department at Chatham House, where she has worked since 2004. Her research areas have included access to energy in low-income countries, sustainable transitions in oil- and gas-exporting economies, and transboundary water relations in western Asia (the Middle East). Since 2009, she has worked on recommendations for energy and water policy and efficiency with partners in the Arab Gulf countries. She is currently leading an international initiative called Valuing Vital Resources on the costs and price linkages between energy, water and food and is on the organizing committee of the Moving Energy Initiative.

From 2002 to 2004, Glada was a senior research fellow at the Gulf Centre for Strategic Studies and has since worked for a number of organizations as a freelance consultant.

She has a BA in Arabic and international relations, including a year spent at the University of Damascus between 1999 and 2000, and an MA in Near and Middle Eastern studies from the School of Oriental and African Studies, London (2008). Glada was appointed as a trustee to the board of Capoeira4Refugees, working with vulnerable children in Jordan, Syria and Palestine, in 2013.

**Owen Grafham** joined Chatham House in May 2014 as a programme coordinator within the Energy, Environment and Resources Department. During his time in the department, he has coordinated Chatham House's research and outreach on energy for displaced populations and the institute's drive to improve the way that energy is integrated in the global humanitarian response. He also coordinates the department's 'Fossil Fuel Expert Roundtable' series and its work on the petroleum governance of emerging oil and gas producers.

Prior to joining Chatham House, Owen spent two years at the Institute for Strategic Dialogue (ISD), where he managed a project that sought to establish concrete areas of trilateral cooperation between the United States, Turkey and Europe. From 2008 to 2010, he worked in Hong Kong with the Hong Kong Institute for Education (HKIEd) and in Sudan, primarily with the Sudanese Development Initiative (SUDIA).

Owen has a BA in politics and literature from the University of York and an MSc in African politics from the School of Oriental and African Studies, London.

### Contributing authors and data modelling

**Dr William Blyth** has worked for 20 years in the analysis of energy security and climate change. He is an associate fellow of Chatham House and an honorary research fellow at Imperial College London. He is director of Oxford Energy Associates, an independent energy research company. He is an expert in energy sector economics, policy and modelling, specializing in transitions to a low-carbon economy and energy access in developing countries.

**Johanna Lehne** is a research assistant in the Energy, Environment and Resources department at Chatham House, working predominantly on energy for displaced populations and China's role in global resource governance. She has a BA in European social and political studies from University College London and an MPhil in international relations from the University of Oxford.

# Acknowledgments


The authors are grateful to the many organizations and individuals that have assisted us during the course of researching and writing this report.

Thanks are firstly due to the Humanitarian Innovation and Evidence Programme at the UK Department for International Development (DFID) for its financial support for the Moving Energy Initiative (MEI). In particular, Razi Latif and Sarah Lester have guided the project from its inception, and contributed their time and expertise to stakeholder discussions.

We would like to thank our consortium partners in the MEI. These include Ben Good, Stephen Okello, Kavita Rai and Lindsay Van Landeghem at GVEP International, whose work in the Dadaab refugee camps (Kenya), in Irbid (Jordan) and on private-sector engagement in the humanitarian sector has been crucial to this report. Similarly, thanks to Mattia Vianello and Mary Wilcox of Practical Action, whose research and work on Goudoubo refugee camp in Burkina Faso and on cooking technologies has been tremendously helpful. Joseph Attwood and Annika Hampson at the Norwegian Refugee Council (NRC) have been a mine of information and made our research in Irbid possible. We are grateful to colleagues at the UNHCR for their help and support in assisting our many avenues of enquiry. Without the transparency and desire for improvement that they have shown, little of this report would have been possible. We are indebted to Betsy Lippman, Amare Egziabher, Paul McCallion, Sam Perkins, Theresa Beltramo and Paul Quigley, all of whom have also contributed to this report in one form or another.

We wish to offer special thanks to all the field staff at the UNHCR, Mercy Corps, the NRC, the Border Consortium, the International Committee of the Red Cross (ICRC), the United Nations Office for Project Services (UNOPS), the World Food Programme, UNICEF, Care International and others interviewed during the course of this project, many of whom chose to speak anonymously and who devoted precious time to contribute to this study, often in harrowing conditions.

The authors would like to thank all those who provided feedback on various drafts of this report. In particular, we appreciate the considered responses of our blind reviewers and Katherine Arnold and Corinne Hart at the Global Alliance for Clean Cookstoves (GACC). We are also grateful to those who gave technical reviews of our data model, notably Ranyee Chiang and her team at GACC, and Kat Harrison and the team at SolarAid.

We would like to thank all the participants at a workshop on 18 June 2015 in London – 'An Unsustainable Situation: Energy for Refugees and Displaced People' – which helped generate and clarify ideas around these issues. We would also like to thank all those involved in the SAFE Humanitarian Working Group for inviting us into the group and attempting to amplify our messages wherever possible.

Finally, thanks are due to our colleagues at Chatham House who have been involved in guiding and assisting this project: Michael Keating, the driving force behind the MEI; Rob Bailey, for his oversight and advice throughout; Johanna Lehne for her tireless research and editorial assistance; Will Blyth for constructing and making numerous revisions to our data model; and to Alice Rowsome and Rosina Norton for supporting the research and data-building processes in the early stages of this project. Thanks also to Gabriela Flores for her input and advice in communicating the report's narrative, and to Elisabeth Jeffries and Jake Statham for copy-editing the report and overseeing the production process at Chatham House.

Any errors or omissions are the responsibility of the authors.

# Acronyms and Abbreviations

| | |
|---|---|
| $CO_2$ | carbon dioxide |
| GACC | Global Alliance for Clean Cookstoves |
| GTF | Global Tracking Framework |
| IASC | Inter-Agency Standing Committee |
| ICRC | International Committee of the Red Cross |
| IDP | internally displaced person |
| IEA | International Energy Agency |
| IFRC | International Federation of Red Cross and Red Crescent Societies |
| IOM | International Organization for Migration |
| IPP | independent power producer |
| kg | kilogramme |
| kgoe | kilogrammes of oil equivalent |
| kWh | kilowatt-hour |
| LPG | liquefied petroleum gas |
| MSF | Médecins Sans Frontières |
| MEI | Moving Energy Initiative |
| MW | megawatt |
| NGO | non-governmental organization |
| NRC | Norwegian Refugee Council |
| PV | photovoltaic |
| SAFE | Safe Access to Fuel and Energy |
| SDG | Sustainable Development Goal |
| SE4All | Sustainable Energy for All |
| $tCO_2$ | tonne of carbon dioxide |
| UNDESA | United Nations Department of Economic and Social Affairs |
| UNEP | United Nations Environment Programme |
| UNHCR | United Nations High Commissioner for Refugees |
| UNRWA | United Nations Relief and Works Agency for Palestine Refugees in the Near East |
| WASH | water, sanitation and hygiene |
| WFP | World Food Programme |
| WHO | World Health Organization |
| WRC | Women's Refugee Commission |

Note: All dollar amounts in this report are in US dollars.

# Preface

The Moving Energy Initiative (MEI) is a collaboration between GVEP International, Chatham House, Practical Action Consulting, the Norwegian Refugee Council (NRC) and the United Nations High Commissioner for Refugees (UNHCR). The initiative began in January 2015, supported by the UK Department for International Development (DFID). The MEI aims to offer solutions for delivering energy in situations of forced displacement in a manner that reduces costs, is safe, healthy and respectful, and also benefits host countries and communities. Where possible it aims to create opportunities for income generation and knowledge transfer to tackle energy poverty and improve energy sustainability.

The first phase set out to raise the level of knowledge about the current energy situation in contexts of displacement globally through desk and field research. The authors canvassed a wide range of stakeholder views in order to assess the extent of the problem and identify challenges and potential approaches. Over the coming phases of the project, the MEI plans to continue generating momentum for change on a global level and promote a 'learning by doing' approach through pilot projects in Jordan, Kenya and Burkina Faso. These local activities will aim to demonstrate new approaches on the ground, and will be geared towards delivering practical improvements in sustainable energy access for refugee and host communities.

The project is grateful to the UNHCR for its openness and transparency in sharing data and facilitating access to sites, staff, populations of concern and partners. Because of this help, this report often uses the UNHCR as a reference point for its analysis and recommendations. However, the report's findings are not based exclusively on observation of UNHCR activity; they also reflect research within the wider humanitarian system. The challenges faced by the UNHCR in reducing energy poverty are comparable to those faced by other humanitarian agencies. We recognize that the UNHCR and many other humanitarian organizations are in the process of evaluating energy practices and revising policy and management systems to reduce their carbon footprint.

This report has been produced by Chatham House on behalf of the MEI. The report draws on consultations within the MEI consortium, but ultimately reflects the authors' views. It does not necessarily reflect all the views of the consortium's members, and any mistakes or omissions are the responsibility of the authors alone. The report is based partly on an evolving and continuously updated dataset, which uses a purpose-built model developed by Chatham House to estimate energy use and $CO_2$ emissions among displaced households. This means that the data cited in this report, and its underlying analysis, may be adjusted in line with new information as the project progresses. Accompanying 'toolkits' – practical guides offering greater depth on specific aspects of energy provision and sustainability – referenced throughout this report will be available online on the MEI webpage – https://www.chathamhouse.org/movingenergy.

#MovingEnergy

# Foreword

There are now 60 million forcibly displaced people on our planet – more than the population of Australia and Canada combined. They include refugees, asylum-seekers and internally displaced persons (IDPs).

This numbing figure is likely to increase further unless concerted action is taken to address the root causes of violent conflict. At a time when the humanitarian system is overstretched and underfunded, nothing could be more urgent.

In the meantime, the imperative is to find humane, creative and cost-effective ways to respond to the needs of so many individuals, most of whom are women and children.

Improving access to clean, safe and sustainable energy offers a promising way forward.

Everybody needs energy services for light, heat, cooling, communication and mobility. However, as the MEI highlights, the costs of energy access and provision are unnecessarily high, whether measured in terms of finance, the environment, health or security.

Entrepreneurship and amazing advances in technology are not being used systematically to respond to the needs of uprooted people or the communities that host them.

Getting this right could yield significant benefits for humanitarian organizations, host authorities and governments and above all for the livelihoods and dignity of the forcibly displaced.

**Kofi Annan**

# Executive Summary

Displacement of people as a result of conflict is not a new phenomenon – but today it represents an unprecedented global challenge. The gap between the needs of growing numbers of displaced people and the resources and political will to meet their needs is widening. For example, voluntary contributions met less than half the $3.05 billion increase in the UNHCR's funding requirement between 2009 and 2013.

Energy is one critical area which illustrates this problem but also offers potential for practical redress. Energy services are essential for basic human protection and dignity, two of the core ethical aims of humanitarian assistance. Energy services provide cooking, lighting, heating and clean water, and underpin all but the most rudimentary income-earning activities. Yet millions of displaced people lack access to clean, safe and secure energy services, in part because funding for such services is inadequate. The lack of reliable data on energy use in the humanitarian field shows that it is a neglected area. But the evidence amassed in the course of this project reveals a huge opportunity to provide better and more sustainable energy services.

Drawing on open-source data, interviews and field surveys, this report offers the first global overview of the state of energy use among almost 60 million people forcibly displaced by conflict.[1] It considers the mounting financial and human costs of their current methods of obtaining energy, and assesses the economic, environmental and human case for change.

## Key findings

**1. Energy use by displaced people is economically, environmentally and socially unsustainable. Children and women bear the greatest costs.**

**Few forcibly displaced people have access to modern forms of energy, yet this group is not represented in international initiatives to improve energy access.** Preliminary calculations indicate that 80 per cent of the 8.7 million refugees and displaced people in camps have absolutely minimal access to energy, with high dependence on traditional biomass for cooking and no access to electricity. This state corresponds with 'Tier 0' in the Sustainable Energy for All (SE4All) initiative's Global Tracking Framework (GTF) for improving energy access worldwide. All SE4All's partner countries and organizations support wider access to energy, and this is now enshrined in Goal 7 of the Sustainable Development Goals (SDGs). However, there is as yet no explicit consideration of displaced people in SE4All's agenda, the SDGs or most countries' energy access targets.

**In 2014 household energy use among forcibly displaced people amounted to around 3.5 million tonnes of oil equivalent, predominantly in the form of firewood and charcoal.** The cost of this fuel is not easy to estimate: sometimes biomass will be collected for 'free'; in other cases fuel prices are much higher in remote camps than for the general population. Conservative estimates suggest expenditure would be at least $200 per year per family of five, which works out at a global total of $2.1 billion per year. That cost is paid chiefly by displaced people, with some of the expense supplemented (often at a much higher cost per unit) by humanitarian agencies and host governments.

**Minimal energy use generates disproportionate emissions.** At around 13 million tonnes of carbon dioxide ($tCO_2$) a year, estimated emissions from displaced households' energy use represent a small proportion of global emissions. However, inefficient burning of biomass means that such emissions are high relative to the energy consumed. Firewood consumption emits 4.54 $tCO_2$ per tonne of oil equivalent, compared with 2.79 $tCO_2$ from burning an equivalent amount of liquefied petroleum gas (LPG).

**Human health suffers as a result of inadequate energy services.** This report estimates, based on World Health Organization (WHO) data, that dependency on primitive fuels is a cause of premature death for some 20,000 displaced people each year as well as respiratory and heart conditions affecting children and the elderly.[2] Open fires, kerosene lamps and candles are all common causes of fires, especially in dry climates or where shelters are made of wood and textile. Women and girls frequently experience intimidation and sexual violence when leaving camps to collect firewood. Children are sometimes poisoned by accidentally drinking kerosene from plastic bottles.

[1] This report considers the energy situation for all people displaced by conflict, estimated at 59.5 million for 2014. However, for statistical purposes, it draws on a data model custom-built by Chatham House which is based on 49.05 million of the 'persons of concern to the UNHCR', as listed in the statistical annexes to the *UNHCR Global Trends 2014: World at War*, Annexes, http://www.unhcr.org/pages/49c3646c4d6.html.

[2] This back-of-the-envelope calculation first took the ratio of deaths as a result of indoor air pollution – 4.3 million people annually as estimated by the WHO – to the total number of people dependent on solid biomass globally – 2.9 billion as estimated by the World Bank. This ratio was then applied to the number of displaced people we estimate to be reliant on solid biomass. Better studies of pollution-related health issues in situations of displacement would be needed to gain a more accurate estimate.

**An estimated 64,700 acres of forest (equivalent to 49,000 football pitches) are burned each year by forcibly displaced families living in camps.** Deforestation is a common problem around refugee camps. Costs and security risks increase as families are forced ever further afield in search of firewood in the absence of alternative sources of fuel.

**Most refugee camps are reliant on poorly planned, inefficient diesel solutions to power offices, schools, hospitals and community facilities.** A lack of reliable data and major differences between camp operations make estimating diesel costs difficult. However, case studies show that, for example, approximately $2.3 million a year is spent on diesel in the Dadaab refugee camps (established in 1992) in Kenya. If a similar amount of fuel were spent relative to the number of camp inhabitants worldwide, it would cost the UNHCR around $56 million a year. Transport costs for staff and equipment are additional to this and largely unaccounted for by humanitarian agencies.

### 2. Improving access to cleaner and more modern energy solutions would reduce costs, cut emissions and save lives.

**The widespread introduction of improved cookstoves and basic solar lanterns could save $323 million a year in fuel costs in return for a one-time capital investment of $335 million for the equipment.** In each case, substantial maintenance, training and support costs would be necessary to make such an intervention effective and durable. The annual fuel saving would mainly accrue to displaced people, who currently spend substantial proportions of meagre household incomes on energy. Such an intervention could also result in emissions savings of around 6.85 million $tCO_2$ per year.

**Use of the best available technologies for household energy services could save 11.38 million $tCO_2$ in emissions each year and bring greater human and environmental benefits.** Widespread introduction of LPG or biogas cookstoves and solar photovoltaic (PV) mini-grids could transform the lives of displaced people and help reduce deforestation.

### 3. The barriers to a sustainable, healthier, more cost-effective system are not technological but institutional, operational and political.

**There is a severe shortage of energy expertise in the humanitarian system and no systematic approach to planning for and managing energy provision.** The design of energy solutions is technical, complex and highly dependent on context. The humanitarian system lacks dedicated energy experts with the requisite skills and knowledge.

**Short-term, politically oriented humanitarian funding is poorly suited to financing longer-term energy solutions in protracted crises and recovery situations.** Humanitarian agency planning and budgets are generally annual, with few incentives to make longer-term investments. No formal cluster of agencies is responsible for energy provision in emergencies, in contrast to other basic needs such as food, water, shelter and health. As a result, donors are not presented with energy as a strategic priority. This restricts funding opportunities, and impairs energy programme prioritization and coordination.

**Political sensitivities prevent rational approaches.** According to the UNHCR, the average amount of time spent as a refugee is 17 years. This can be an uncomfortable truth for host governments dealing with local resentment towards refugees. It may make politicians reluctant to endorse medium- to long-term energy investments that imply some degree of permanence for refugee populations. Yet these are the very investments that provide optimal energy solutions.

**The humanitarian sector's 'procure and provide' model precludes opportunities for better energy services.** The skills shortage among implementing agencies and a tendency for short-termism among donors and host governments perpetuate a 'procure and provide' model for energy equipment distribution among displaced populations. For cleaner energy options, agencies frequently rely on equipment donations with little consideration of local context or end user preferences. Too often, success has been measured by the number of products distributed, such as solar lamps or efficient cookstoves, rather than by their impact. Attempts to increase cleaner energy access frequently stop at the pilot stage, missing opportunities both to bring down costs through scaling up demand and to develop appropriate household payment models.

### 4. Doing things differently can bring significant benefits for host countries.

**Sustainable energy solutions reduce environmental and social pressures and create opportunities for local businesses.** By curbing firewood demand, clean cooking technologies can reduce environmental degradation and related resource tensions with local communities. In some cases, there may be opportunities for local energy service companies to help meet the needs of displaced populations.

**Energy investments help integrate displaced populations and provide a legacy asset for local communities.** Relevant approaches are being piloted in Jordan. For example, the Norwegian Refugee Council is installing solar panels in schools receiving Syrian children, as well as solar water heaters in residential buildings in return for guaranteed periods of accommodation at reduced rents for refugees. The UNHCR is funding a solar farm with several partners outside the camp of Azraq as a legacy for the country after the refugees leave.

**Sustainable energy solutions can contribute to national and local sustainable development objectives.** Governments of countries hosting displaced people will have policies or ambitions to reduce carbon emissions and scale up efficiency and renewable energy. Many governments want to tackle deforestation. This presents opportunities to collaborate with donors and implementing agencies on energy solutions that both meet the needs of displaced persons and respond to national sustainable development priorities.

## Six imperatives for change

Changing approaches to energy supply for forcibly displaced populations can cut costs, reduce environmental impacts and save lives. It can also bring important benefits to local communities and national populations in host countries. This report identifies six imperatives for change:

### 1. Incorporate sustainable energy access for displaced people into international, national and agency agendas.

At multilateral and national government levels, this means integrating the issue of sustainable energy for the forcibly displaced into the UN-led SE4All agenda, and developing an action agenda specifically for displaced people within the post-2015 SDGs. At agency level, it means incorporating energy considerations into 'core programming' – that is to say, the basic operations and procedures of humanitarian agencies – at each stage of the humanitarian response. Creating regular positions for renewable energy experts would make this task easier.

### 2. Build the data.

All relevant agencies should collect detailed energy-related data for refugee camps and other displacement contexts, and use standardized methods for data reporting. The data should cover energy use, costs, supply and transportation fees, and equipment efficiencies involved in both (a) energy use by displaced populations; and (b) energy provision for camp facilities and humanitarian operations. Assessments of local entrepreneurship models, as well as of displaced households' income and spending, should be included. Such a process will help to inform the cost–benefit evaluations needed for capital investment in energy, and provide a basis for competitive tendering.

### 3. Coordinate national ambitions and humanitarian aims for mutual benefit.

If forcibly displaced people are unlikely to return to their country of origin within a short time frame, energy interventions should be coordinated with local and national government authorities. Countries hosting refugees have ambitions to increase the sustainability of their energy systems and often to increase energy access for their own populations. Energy interventions will have the greatest chance of being accepted and supported if they aim to support these national goals. As part of this conversation, host governments and agencies may need to discuss the lifting of restrictions on displaced people's rights to work and access to land, as these may inhibit their ability to pay for and access energy services.

### 4. Embed energy projects and accountability at the local level.

Longer-term solutions are viable only if host populations and governments support them. This requires understanding of the needs of local communities, and of the economic linkages between such communities and displaced populations. A good understanding is also needed of local laws and regulations, the capacities of local energy service providers, the nature of local energy markets, and geographic and climatic factors affecting technology choice. Accountability for the performance of energy projects must lie with stakeholders on the ground, and must be long-term. Energy providers and appointed camp regulators are examples of the types of body that might have such a remit.

### 5. Explore new delivery models.

Initial emergency relief should move towards more sustainable energy provision based on the self-reliance of refugees and internally displaced persons (IDPs). This means working out how displaced people can access and pay for energy services, and how private-sector expertise can be leveraged through innovative tendering and private–public partnerships. An overhaul of energy service procurement policy and standards is required to engage a wider cast of private-sector actors in energy service delivery and equipment sales. Contracts should be carefully designed to incentivize efficiency and sustainability while ensuring humanitarian aims are met.

### Practical ways forward

- Establish a Moving Energy Initiative (MEI) advisory panel of humanitarian agency leaders, political leaders, and technical and financial experts to steer and promote implementation of the MEI's recommendations. The aim would be to build support and funding for scaling up successful energy delivery solutions worldwide.
- Create a revolving fund that can lend to agencies proposing to invest in energy service projects. An inter-agency trust will be needed to manage this central pool of resources. This should have the expertise to advise on the contracting and regulatory frameworks required to deploy funds effectively and accountably.
- Revise models for camp planning with sustainable energy objectives in mind. These models should be widely shared among government, humanitarian and other relevant partners to assist with energy and related decision-making. The aim would be to set up camps in such a way as to avoid locking them into inefficient and inappropriate energy models.
- Establish an energy dialogue between the private sector and humanitarian organizations to develop and harmonize comprehensive and progressive technology standards.
- Pilot site-specific integrated energy plans in several large displacement zones, with monitoring and evaluation over several years.
- Explore and pilot the implementation of concessions for meeting cooking needs at scale without wood or charcoal. These need to both bring down overall costs and significantly reduce or eliminate wood reliance in each camp/area over a period of years.

### 6. Explore innovative funding models.

Donors should allow their funding to be deployed in forms that can 'de-risk' private-sector investment and kick-start local markets and supply chains. Solutions based on local markets can create opportunities for income generation by drawing on the entrepreneurial talent of displaced people and host communities. Cash transfers to vulnerable households may prove more effective than fuel handouts, by allowing choice over energy services. Energy service contracts for camp facilities may include 'bolt-on' obligations to expand electricity access to households.

These reforms will not be straightforward. They require investment, a long-term perspective and a willingness to innovate and experiment, all of which may be absent when money is short and the immediate priority is to save lives. The burden of change falls not only on humanitarian agencies, but also on donors to encourage reform and on host governments to back new approaches.

Yet the conditions for reform have never been better. Change is already under way in the humanitarian sector. This report shows how agencies are experimenting with new technologies and delivery models. Some larger agencies have established the SAFE ('Safe Access to Fuel and Energy') Humanitarian Working Group, which acts as an international coordination mechanism for humanitarian energy response and works to improve energy delivery to crisis-hit populations. The forthcoming World Humanitarian Summit in 2016 provides a historic opportunity to galvanize the international community into action and scale up existing efforts.

Rapid change is also taking place outside the humanitarian system. Technological advances offer continual opportunity for improvement. Falling prices for technologies like solar PV and light-emitting diodes mean sustainable energy solutions are now more cost-effective than traditional technologies in many countries. This is particularly relevant to off-grid rural areas, where it can be prohibitively expensive to extend the electricity grid and where transportation adds to fuel costs. Meanwhile pay-as-you go financing models using mobile phones and smart metering are proving successful in enabling electricity access in many parts of Africa and Asia.

Across the planet, wider access to clean energy is a rising priority. This is crystallized in the new SDGs, the seventh of which commits the world to 'affordable, reliable, sustainable and modern energy for all' by 2030. With new technologies and delivery methods, the humanitarian agencies can assist in meeting this target while also saving lives and offering livelihood and development opportunities. Given the current state of energy provision, doing nothing is a betrayal of humanitarian principles.

# 1. Introduction: the Energy Problem

The number of people across the world forced to leave their homes and seek refuge elsewhere due to conflict is approaching 60 million.[3] This includes refugees and internally displaced persons (IDPs), referred to together in this report as 'displaced'.[4] This number has doubled in the past decade and is greater than the populations of Australia and Canada combined. This creates many practical and policy challenges. Governments, humanitarian agencies, and a vast array of NGOs and local communities are involved in accepting and taking care of displaced people. Yet from East Aceh in Indonesia to Calais, France, it is clear that needs are overwhelming the assistance available.

Saving lives, protecting vulnerable people and maintaining human dignity are all core ethical aims of humanitarian assistance. It is the duty of countries in which people seek refuge from war or persecution to fulfil these aims according to the United Nations' 1951 Convention Relating to the Status of Refugees.[5] This study considers one aspect of this response that is critical for achieving each of these aims yet strikingly under-resourced: energy. It examines the mounting financial and human costs of energy supply and use in displacement situations, and urges change.

In short, the evidence shows that energy provision in its present form undermines the fundamental humanitarian aims of assistance. A different approach aiming for more sustainable and cleaner energy delivery could provide multiple benefits both to displaced people and to the countries that host them.

**Figure 1: Total population of concern to the UNHCR,[6] 2004–14 (million)**

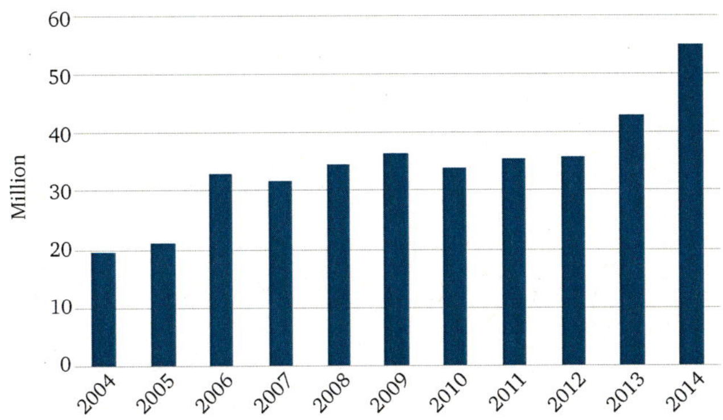


Source: *UNHCR Global Trends 2014: World at War,* Annex, http://www.unhcr.org/pages/49c3646c4d6.html.

---

[3] According to the UNHCR, there were 59.5 million forcibly displaced people worldwide in 2014. *UNHCR Global Trends 2014: World at War*, http://www.unhcr.org/pages/49c3646c4d6.html.

[4] People who have had to leave their homes as a result of an event related to conflict. Includes refugees, asylum-seekers and IDPs (including IDPs beyond the UNHCR mandate).

[5] UNHCR, *The 1951 Convention Relating to the Status of Refugees and its 1967 Protocol.*

[6] The categorization of the 54.9 million 'persons of the concern to the UNHCR' is not synonymous with the 59.5 million forcibly displaced people globally. It includes refugees, asylum-seekers, IDPs, returnees and stateless persons under the UNHCR mandate. It excludes Palestinian refugees, who are under the remit of UNRWA.

## A deepening crisis

Number of forcibly displaced persons requiring UNHCR help.

## Women bear the greatest costs

Women and girls frequently experience intimidation and violence when collecting firewood. Some 500 displaced Darfuri women and girls were raped while collecting firewood and water within a five-month period in Sudan (Médecins Sans Frontières, 2005).

### Box 1: What are energy services?

Power, light, heat, cooling and mobility are all energy services. They underpin almost all aspects of human security and well-being. In its various forms, energy supports good health and nutrition. Its benefits include liveable temperatures, water treatment, cooked food and medical facilities. Energy provides lighting for self-education, schools and safer streets. It is needed for community facilities, mobile phone chargers, radio and television, all of which facilitate human contact. Power, communications, lighting and mobility also provide opportunities for people to earn a living.

Energy services can be supplied in many different ways. These include, at the most basic level, physical human or animal effort. They also encompass applications that capture renewable energy sources, such as solar radiation, wind and hydropower. However, the majority of energy services worldwide are provided by combustion of fuel – chiefly oil, coal, gas and biomass that includes wood. Electricity is a cleaner way to provide cooking/heating and lighting services than direct fuel combustion. It can be generated either through renewable sources or through fuel combustion in electricity generation plants.

Many studies emphasize the particular importance of access to so-called 'modern energy services' for women[a] and children.[b] The term encapsulates reliable access to electricity as well as to clean – i.e. safer and relatively non-polluting – cooking facilities thanks to the use of cleaner fuels and more efficient appliances.[c] Access to affordable and reliable energy services is a recognized development multiplier and is now listed as one of the global Sustainable Development Goals for 2015–30.

[a] See, for example, Women's Refugee Commission (WRC), 'Protecting Women and the Environment in the Great Lakes Region', https://womensrefugeecommission.org/resources/document/1132-safe-energy-great-lakes-2015.
[b] Hannah Strohmeier, *Why sustainable energy matters to children: The critical importance of sustainable energy for children and future generations*, UNICEF, 2015, http://www.unicef.org/environment/files/UNICEF_Sustainable_Energy_for_Children_2015.pdf.
[c] International Energy Agency (IEA), World Energy Outlook, 'Modern energy for all: why it matters', http://www.worldenergyoutlook.org/resources/energydevelopment/modernenergyforallwhyitmatters/, last accessed on 19 October 2015.

## A unique set of conditions facing displaced people

Energy services (see Box 1) are of critical importance to displaced people, many of whom live in temporary shelters exposed to temperature extremes. Medical care requirements, separation from relatives and lack of income and legal status all affect the need for, use of and ability to access services. Most settlements of displaced people are not connected to gas and power supplies, and are far from or on the outskirts of urban centres.

**Medical care requirements, separation from relatives and lack of income and legal status all affect the need for, use of and ability to access services.**

Paying for energy takes up a significant share of the low and insecure incomes of the many displaced people living among host communities. As with water and food, energy delivery usually affects local economies and living environments. Many countries with large numbers of displaced people already suffer from wider resource stress, manifest for example in deforestation and energy poverty. Thus additional competition for fuel can exacerbate tensions between local and displaced communities.[7]

Energy provision has particular characteristics and so is worth considering separately from provision of other humanitarian services such as water, sanitation, health, food and shelter. But it is also connected to each of these areas, and in some respects presents comparable challenges. Good energy management necessarily requires the expertise, practices, materials and governance involved in delivering other humanitarian services.

## Growing numbers, tightening budgets

The developing world now hosts the majority of refugees (86 per cent).[8] Around 82 per cent of persons of concern to the UNHCR[9] live outside camps – in rented accommodation, informal settlements or even out in the open.[10]

[7] Adrian Martin, 'Environmental Conflict Between Refugee and Host Communities', *Journal of Peace Research*, Vol. 42, No. 2, May 2005, pp. 329–46.
[8] *UNHCR Global Trends 2014: World at War*.
[9] See Appendix B for a definition of persons of concern to UNHCR.
[10] 82 per cent is the non-camp share of the population of concern to the UNHCR based on authors' analysis of the statistical annexes to *UNHCR Global Trends 2014: World at War*. If numbers from the UNRWA are added, the share is 84 per cent; 29 per cent of Palestinian refugees registered by the UNRWA live inside camps, and the conditions in these camps are similar to those of the urban poor in other developing countries. See UNRWA, 'In Figures', www.unrwa.org/sites/default/files/2014_01_uif_-_english.pdf.

It is difficult to forecast the number of displaced people, but several factors suggest it will continue to rise. Repatriation in the next few years will not be an option for the majority due to politics and conflict in their places of origin.[11] More generally, climate change, environmental stress and poor resource governance are likely to exacerbate the conflicts that cause displacement globally.[12] The need for services to support displaced people, many of whom lack citizenship rights and income, will grow accordingly. To compound the problem, the very energy used to provide these services is chiefly created using inefficiently burned, unhealthy and expensive biomass or fossil fuels.

In most cases, humanitarian relief efforts are insufficient. Humanitarian agencies often face budget constraints, which means that funds are prioritized for saving lives in emergencies and that 'protracted caseloads' are systematically underfunded. The problem of underfunding is illustrated in Figure 2, which displays UNHCR funds requested in 2014 against those received.

**Figure 2: Gap between funds requested by UNHCR and funds received in 2014 ($ million)**

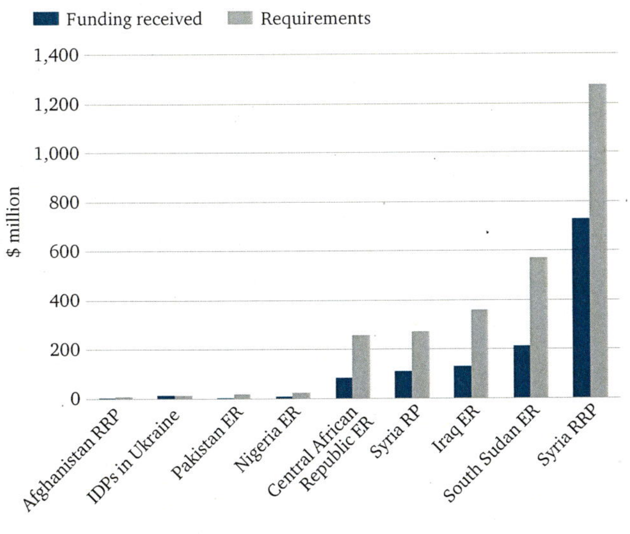


RRP = Refugee response plan
ER = Emergency response
RP = Response plan

Source: *UNHCR Global Appeal 2015 Update*, 'Identifying Needs and Funding Requirements', http://www.unhcr.org/5461e5f30.html.

## Temporary solutions maintained for years

Humanitarian responses to large-scale forced migration require the emergency supply of water, food, shelter, cooking and lighting equipment, and facilities such as schools, clinics and administration offices. The uncertain and often controversial status of refugees means that they must be treated as temporary residents, but in practice energy services are usually extended for many years in an ad hoc and inadequate manner. Indeed the average length of time spent as a refugee is around 17 years.[13] In most cases, household energy use is the responsibility of displaced people. As a result of these conditions, they usually pay above market prices for heating and lighting, and take physical risks in obtaining and using fuel.

**The uncertain and often controversial status of refugees means that they must be treated as temporary residents, but in practice energy services are usually extended for many years in an ad hoc and inadequate manner.**

The lack of resources, long-term planning and financing dedicated to energy services means that opportunities are being missed to improve the quality of life both of displaced people and of local populations.

## A gap in the international energy access agenda

Energy access and the need to shift to cleaner energy service provision are now priorities for donor countries, multilateral agencies and financiers involved in helping displaced people. As Box 2 shows, there is commitment on a global level to tackling these issues. Many countries hosting refugees are already dealing with severe energy access challenges, deforestation, high fuel costs and fuel pollution. Yet somehow displaced people fall through the net. Despite their specific needs and conditions, this group is not considered explicitly in the SE4All initiative or the SDGs. We have not come across any host government that specifically incorporates displaced people in its policies on sustainable energy or energy access.

[11] For example, see Katy Long, *The Point of No Return: Refugees, Rights, and Repatriation*, Oxford University Press, 2013.
[12] Jürgen Scheffran and Antonella Battaglini, 'Climate and conflicts: the security risks of global warming', *Regional Environmental Change*, 2010, 11:1, pp. 27–39.
[13] UNHCR, 'Protracted Refugee Situations', Executive Committee of the High Commissioner's Programme, Standing Committee, 30th Meeting, p. 2, http://www.unhcr.org/40c982172.pdf.

### Box 2: Global commitment to sustainable energy access

Scientific evidence shows that if global warming is to be limited to an average of 2°C above pre-industrial levels, then emissions of greenhouse gases including $CO_2$ must fall rapidly from 2020, with society moving towards carbon neutrality by mid-century.[a] Global negotiations on climate change thus emphasize the urgency of decarbonizing energy systems and halting deforestation.

This presents a dual challenge for low income countries. It is widely acknowledged that these countries have a moral right to increase their emissions in line with development aims and that energy access is closely linked to reducing poverty. Meanwhile, their populations should be able to access cleaner, sustainable energy and not have to suffer from dangerous levels of pollution. The WHO estimates that 4.3 million people die each year due to indoor air pollution, chiefly from burning biomass and hydrocarbon fuels.[b] These are among the 2.9 billion poorest people in the world, which includes a large proportion of displaced people.

The SE4All initiative, whose partners include all the major development banks, the OPEC Fund for International Development and several UN bodies, recognizes that energy services help eradicate poverty. It aims to make 'cleaner and more efficient' energy available to all by 2030.[c] Major donors, including Norway, Japan, the UK, the EU and the US, are all committed to the above principles. In order to practise what it preaches, the UN itself has pledged to reduce its environmental footprint through its Climate Neutral Strategy.[d]

The SDGs, which most governments are expected to endorse in 2015–16, reinforce this commitment since four of the 17 goals are relevant to sustainable energy access. These encompass ensuring healthy lives and promoting well-being for all (Goal 3); ensuring access to affordable, reliable, sustainable and modern energy for all (Goal 7); ensuring sustainable consumption and production patterns (Goal 12); and sustainably managing forests (part of Goal 15). The SDGs are the result of a global consensus-building process (Rio+20 UN Conference on Sustainable Development, Earth Summit, Rio Conference, Rio Declaration) and will be in force until 2030.

[a] UN Environment Programme (UNEP), *The Emissions Gap Report 2014: A UNEP Synthesis Report*, November 2014, http://www.unep.org/publications/ebooks/emissionsgapreport2014/portals/50268/pdf/EGR2014_LOWRES.pdf.
[b] WHO, 'Indoor air pollution', http://www.who.int/indoorair/en/.
[c] See http://www.se4all.org/.
[d] UNEP, *Moving Towards a Climate Neutral UN: The UN System's Footprint and Efforts to Reduce It, 2014 Edition*, February 2015, http://www.greeningtheblue.org/sites/default/files/brochure_sequential_0.pdf.

## Scope of the study

This report sets out the findings from six months of initial research on the volume and cost of fuel used by forcibly displaced people and the institutions that serve them. This is a diverse and complex subject with little or no open-source data available. In order to better understand and evaluate energy use by displaced populations, Chatham House developed a data model (see Box 3) focused on household energy consumption. The authors also researched energy use for camp facilities and operations, but the data gathered were not robust enough to scale up to a global-level estimate.

This introduction is the first of six chapters followed by four appendices. Chapter 2 sets out the costs of present energy use in terms of financial outlays, human security, social cohesion and environmental protection. Chapter 3 examines the opportunities for change presented by this cost assessment, as well as by global technology developments and new approaches in the humanitarian system. Chapter 4 highlights some alternative approaches that could prove instructive. Chapter 5 outlines the challenges of implementing sustainable energy solutions given institutional, national and local contexts and the obstacles that these present. Chapter 6 provides conclusions, high-level recommendations and practical suggestions on ways to move forward.

### Box 3: An Overview of the Chatham House Model

For the MEI project Chatham House designed a model offering the first estimates of the scale and cost of energy use and $CO_2$ emissions among the households of populations of concern to the UNHCR worldwide.

The statistical annexes to the UNHCR report, *UNHCR Global Trends 2014: World at War*, provided the data on the location and size of the displaced populations. This dataset covers 49,053,874 displaced people and sorts them by country and settlement type. We made a distinction between the population living in camps and the population living outside camps.

**Camp population**
For displaced households in camps we developed a set of baselines for energy use. For cooking the baseline types were firewood-dependent household, firewood/charcoal mix household, liquid fuel-dependent household, LPG-dependent household and alternative biomass-dependent household. For lighting, the baseline types were torch-dependent, kerosene-dependent, electricity-dependent and solar-dependent.

For each of these baseline household types, we calculated the average fuel consumption, energy cost and $CO_2$ emissions using data drawn from 24 semi-structured interviews with UNHCR staff in different countries and independent research on displacement settings. We then assigned to each camp in the study one baseline type for its energy use for cooking, and one for lighting, from the above categories. This was decided on the basis of interview data, field research by partner organizations and independent desk research.

**Non-camp population**
Energy use for the (much larger) non-camp population in the dataset was determined by location, i.e. urban, rural or slum. Allocating proportions of the population to each of these three categories required several steps. First, we separated the total population into urban and rural cohorts, based on Global Tracking Framework (GTF) data showing urban–rural ratios for each country. We applied a further weighting of our own, to reflect the relatively higher proportion of refugees who live in towns and cities (in effect, the urban ratios we used for each country were somewhat higher than those indicated by the GTF data). We then used UN-Habitat data on the prevalence of slums to allocate a share of the urban refugee population in each country to the 'slum' category.

We used GTF data mapping urban and rural energy access ratios (without distinction between non-displaced and displaced people) to estimate the proportion of the displaced population in each setting with access to grid connections and solid/non-solid fuels respectively. However, for displaced households in slums, we estimated access to grid electricity by simply averaging urban and rural ratios for grid access. To estimate slum households' access to solid/non-solid fuels, we used ratios of urban access to these fuels as proxies.

The next step was to translate these ratios into actual estimates of energy consumption. We did this using country-level International Energy Agency (IEA) data on average energy consumption per household for grid electricity and LPG. (In this instance we assumed non-solid fuel to be LPG.) For households not connected to the grid or without access to LPG, we reverted to the refugee camp baseline types. For urban households without access to non-solid fuels, we assigned baseline Type 2 (firewood/charcoal mix). For slum and rural populations without access to non-solid fuels, Type 1 (firewood-dependent) was assigned. All non-camp populations without grid connection were assigned a Type 1 baseline lighting type (torch-dependent). We applied these consumption estimates to the urban/slum/rural population breakdowns for each country to estimate energy usage for each cohort in each country.

**Global estimates**
We calculated **energy spending** by multiplying the number of displaced households in each camp and non-camp context by the average annual household fuel spend for its respective baseline type or national displacement context. Thus, for example, the number of households categorized as firewood-dependent was multiplied by the average fuel spend for a firewood-dependent household. By adding up each camp's annual fuel spend and (for non-camp settings) each country's fuel spend, we estimated global energy costs in forced displacement situations.

A similar method was used to calculate global estimates for **energy consumption** and **emissions**. For these calculations we multiplied the number of displaced households in each camp and non-camp context by the average annual household fuel consumption (for the consumption calculation) and by the average annual household energy emissions (for the emissions calculation). We then summed up all camp and country energy consumption and emissions figures to derive global energy use and emissions figures respectively.

**Limitations of the model**
This simplified model of a highly complex system has a number of limitations. Our baselines, for example, do not fully cover what is in reality a vast array of different patterns of energy use. Our model relies heavily on basic proxies for energy use within particular countries. Energy use by displaced households is likely to vary both within and between camps, but our model only takes the latter into account. Heating is

not considered explicitly in the model. However, the use of fuel for heating is considered to be largely synonymous with fuel use for cooking. Many displaced people, for example, rely exclusively on the warmth of the cooking fire for heating. Costs of heating fuel are also often beyond the capacity of many displaced populations. In contexts where expenditure on fuel for heating is significant, such as during winter in western Asia and Europe, the use of fuel explicitly for heating is worthy of further investigation. Numerous news stories attest to the severity of being unable to meet this need.[a] The model thus gives only an indicative estimate of the kinds of numbers involved and should not be viewed as a comprehensive picture of energy use among forcibly displaced people.

The data used in the model both for assigning types and for assumptions on cost and consumption are imperfect. Figures cited in interviews were often aggregated and unverified.

The methodology behind the MEI's model was presented and discussed during an international roundtable of experts in London on 18 June 2015. The detailed technical assumptions behind the cooking and lighting assumptions were analysed and verified by independent experts at the Global Alliance for Clean Cookstoves and at SolarAid. (For more details on the assumptions and data used, please refer to the methodology in Appendix A.)

[a] See, for example, *The Guardian*, 'Winter is coming: the new crisis for refugees in Europe,' 2 November 2015, http://www.theguardian.com/world/2015/nov/02/winter-is-coming-the-new-crisis-for-refugees-in-europe.

# 2. How Much Energy is Being Used, at What Cost to Whom?

Given that humanitarian efforts will require billions of dollars in aid each year, aligning humanitarian and development goals is essential. Energy is one area of overlap, since many displaced people face challenges of poverty and energy access similar to those encountered by local populations. However, to assess the potential for alternative ways of delivering energy, one first needs an idea of the scale of the problem, including current levels and patterns of energy spending.

Energy use in situations of displacement is an opaque topic. No comprehensive dataset on energy use exists for refugee camps, informal settlements or individual accommodation.[14] A handful of previous studies have concentrated on one area of energy provision – usually cookstoves, although a few more recent unpublished studies cover more comprehensive household needs.[15] The authors did not find any studies on humanitarian agencies' energy use.[16]

In this section we set out for the first time an assessment of the current costs of energy used by displaced populations. The primary focus is on household energy both within and outside camps. However, this section also considers the costs incurred by refugee camp managers and staff looking after displaced people.

## A diverse picture

The dynamics of supply and demand for energy services to displaced people are diverse and complex. The situation varies between rural and urban settings. UN agencies and those leading humanitarian response share differing levels of responsibility with governments and implementing partners depending on the context.[17] It is hard to compare services available in upper-middle-income countries such as Jordan, Lebanon or Iran with those in low-income countries such as Burkina Faso or Uganda. The country of origin and former standard of living of the displaced people in question also make a difference; for example, Syrian refugees will use energy equipment different from that of IDPs in the Central African Republic.

[14] This is also the case for data on fuel location (Where is fuel collected from? How is the supply chain constructed?) and fuel origin (Who is doing the collecting and the selling?).

[15] See Sandra Haskamp and Oliver J. Haas, 'Baseline assessments and renewable energy feasibility studies in Ethiopia, Jordan and Bangladesh', project for UNHCR, *Output 3: Baseline Survey Ethiopia Dollo Ado*, INTEGRATION environment & energy, 28 April 2015. And UNHCR, 'Light Years Ahead Project: Monitoring & Evaluation System and Baseline Assessment Report – Uganda Country Report', July 2014.

[16] Rebecca Gunning, *The Current State of Sustainable Energy Provision for Displaced Populations: An Analysis*, Chatham House Research Paper, December 2014, https://www.chathamhouse.org/sites/files/chathamhouse/field/field_document/20141201EnergyDisplacedPopulationsGunning.pdf, p. 36.

[17] For example, some governments, such as those of China and Turkey, have chosen to take on a large role in coordinating assistance for forcibly displaced persons.

## Very limited access to modern forms of energy

Out of 8.7 million refugees in camps, only 11% have access to reliable energy sources for lighting (estimate: Moving Energy Initiative).

## Major toll on human lives

An estimated 20,000 forcibly displaced people die prematurely every year as a result of pollution from indoor fires (based on WHO global estimates). Open fires, kerosene lamps and candles all frequently cause accidents.

Table 1 gives an idea of the diversity of energy conditions in camps – even those located within the same country. The political setting is equally varied. Each country and displacement situation has its own history and legislation influencing attitudes towards refugees and their capacity to settle and work.[18] In addition, the level of humanitarian funding varies from country to country and from year to year, depending on political priority, media attention and the duration of a particular refugee situation.

### Table 1: Prevalent forms of energy use and shelter in selected refugee camps

| Camp/name of location | Country | Date established | Population, end of 2014 | Energy use | | Shelter |
|---|---|---|---|---|---|---|
| | | | | Camp operations | Refugee households/ businesses | |
| Abala | Niger | 2012 | 12,938 | Diesel | LPG, firewood | Tents with plastic sheets, local materials used to improve shelters |
| Azraq | Jordan | 2014 | 11,315 | Diesel | LPG | Zinc and steel shelters |
| Bahn | Liberia | 2010 | 5,257 | Diesel | Firewood, kerosene, charcoal | Tents with plastics sheets, some with corrugated zinc |
| Buramino | Ethiopia | 2011 | 39,471 | Diesel | Firewood, some charcoal, kerosene and solar | Emergency tents and bamboo shelters |
| Dadaab | Kenya | 1992 | 356,014 | Diesel | Firewood, batteries, diesel | Mud walls, corrugated zinc, some tent sheets |
| Doro | South Sudan | 2011 | 50,087 | Diesel | Firewood, some charcoal | Plastic tents; some have more permanent shelters (zinc roof and mud walls) |
| Emkulu | Eritrea | 2000 | 2,762 | Grid connection, diesel | Firewood, some grid | 90% tents (wood and mud), 10% permanent structures (hollow cement blocks) |
| Erbil (Basirma, Darashakran, Kawergosk, Qushtapa) | Iraq | 2005 | 114,669 | Grid connection | Grid connection, kerosene | Concrete slab floors with tents made of canvas, some fabricated structures (mobile homes) made of steel frames and sandwich panels, some brick |
| Fau 5, Abuda, Shagarab (1,2,3), Girba, Kilo 26, Um Gargur, Wad Sherifey | Sudan | 1970 | 75,251 | Some grid connection, diesel, firewood | Firewood, charcoal, dung, some LPG, some grid connection | Temporary shelters, tents and huts; huts constructed with wood, straw and mud |
| Gendrassa | South Sudan | 2012 | 17,975 | Diesel | Firewood | Plastic tents; some have more permanent shelters (zinc roof and mud walls) |
| Goudoubo | Burkina Faso | 2012 | 10,327 | Diesel | Firewood, solar | Tents with plastic sheets |
| Kakuma | Kenya | 1992 | 153,959 | Diesel, solar | Firewood, kerosene, ethanol, charcoal, briquettes | Mud walls, corrugated zinc, some tent sheets |
| Kilis | Turkey | 2012 | 62,371 | Grid connection | Grid connection, LPG | Containers |
| Kobe | Ethiopia | 2011 | 39,214 | Diesel | Firewood, some charcoal, kerosene and solar | Emergency tents, plastic sheeting transitional shelters, mud plastering |
| Kutupalong | Bangladesh | 1995 | 13,176 | Grid connection, diesel | Compressed rice husk, kerosene, solar | Bamboo, wood, corrugated iron sheet, plastic sheet (roof) |
| Mae La | Thailand | 1984 | 46,978 | Grid connection, diesel | Charcoal, charcoal briquettes, firewood | Bamboo houses |
| Mbera | Mauritania | 2012 | 48,910 | Diesel | Charcoal, some butane gas and dung | Semi-durable shelters (local materials, resistant) |
| Nakivale | Uganda | 1958 | 66,691 | Grid connection, diesel | Charcoal, firewood, kerosene | Mud walls, some have iron sheet, some plastic sheet roof, reasonable light inside |

[18] See Appendix D for examples.

| Camp/name of location | Country | Date established | Population, end of 2014 | Energy use | | Shelter |
|---|---|---|---|---|---|---|
| | | | | Camp operations | Refugee households/ businesses | |
| PTP | Liberia | 2010 | 15,300 | Diesel | Firewood, charcoal, kerosene | Plastic sheets, some zinc |
| Sag-Nioniogo | Burkina Faso | 1994 | 1,845 | Diesel | Firewood | Tents with plastic sheets |
| Saranan | Pakistan | 1986 | 18,248 | Diesel, electricity | Firewood, LPG | Mud houses |
| Tabareybarey | Niger | 2012 | 8,147 | Diesel | Firewood, kerosene | UNHCR tents; some have been altered using local materials |
| Tongogara | Zimbabwe | 1998 | 4,976 | Grid connection | Firewood (supply erratic), charcoal (preferred), cotton immersed in paraffin oil, gas | Bricks (for those who can afford it), mud bricks, mud mixed with cement, corrugated iron for roofing |
| Zaatari | Jordan | 2012 | 84,773 | Grid connection | LPG, diesel-generated electricity (formerly illegal grid connection) | 66% caravans, 33% tents |

Note: Some populations change rapidly, especially in new camps, and may even change seasonally. Azraq camp, for example, is now listed by the UNHCR as having 25,774 inhabitants as of 19 October 2015.
Sources: Interviews with UNHCR field staff and Border Consortium; UNHCR website; GVEP International and Practical Action site surveys, 2015; SAFE Project Database, http://www.safefuelandenergy.org/where-we-work/search-projects.cfm.

## Energy use by displaced households

Many displaced people do not have access to enough energy to cover their basic needs. In Chad, some 35 per cent of displaced households surveyed reported having to skip meals during the previous week because they did not have enough fuel to cook with.[19] In the same survey, 28 per cent of households reported undercooking meals in the same period for the same reason.[20] These low levels of energy access are also evident in other locations. In Ethiopia's Dollo Ado camps, 28 per cent of households had sold rations to buy cooking fuel while 38 per cent had undercooked food in the week prior to being surveyed.[21] In Ugandan refugee camps about half the households have admitted undercooking food more than twice a week. About 44 per cent report skipping meals once a week, and about 10 per cent of households sell over a quarter of their family food ration to buy cooking fuel.[22]

Our model also reveals that an estimated 89 per cent of people in refugee camps have Tier 0 lighting, while an estimated 80 per cent have Tier 0 cooking facilities. This means that around 7 million refugees in camps have less than four hours of access to electricity and inadequate access to non-electric energy every day.[23]

Using our model (see Box 3), we estimate that forcibly displaced households of concern to the UNHCR had no choice but to burn around 3.5 million tonnes of oil equivalent for cooking and lighting last year.[24] Most of this was in the form of firewood and charcoal. This would equate to spending of around $2.1 billion. See Box 3 and Appendix A for details on the methodology used.

Figures 3 and 4 below show our estimates of energy use and spending for cooking and lighting in each displacement context.

[19] Out of a sample of 673 households surveyed in Kounoungou and Mille camps in Chad. UNHCR, 'Light Years Ahead Project: Monitoring & Evaluation System and Baseline Assessment Report – Chad Country Report', July 2014.
[20] Ibid.
[21] Out of a sample of 494 households surveyed in Bokolmanyo, Melkadida, Kobe, Hilaweyn and Buramino camps in Ethiopia. Sandra Haskamp and Oliver J. Haas, 'Baseline assessments and renewable energy feasibility studies in Ethiopia, Jordan and Bangladesh', project for UNHCR, *Output 3: Baseline Survey Ethiopia Dollo Ado*, INTEGRATION environment & energy, 28 April 2015.
[22] Out of a sample of 702 households surveyed in Navikale camp in Uganda. UNHCR, 'Light Years Ahead Project: Monitoring & Evaluation System and Baseline Assessment Report – Uganda Country Report', July 2014.
[23] Tier 0 access essentially signifies low or no access to energy. Users with Tier 0 access receive less than four hours of access to very low power (less than 3 watts) during the day and less than one hour in the evening. Daily consumption levels do not exceed 12 watt-hours. Users cannot even use or do not even have access to very low-power appliances like solar lanterns. See Gabriela Elizondo Azuela, 'Sustainable Energy for All Global Tracking Framework 2015: Workshop on Capacity Development for Mainstreaming Energy Sustainable Development Goals (SDGs), Targets and Indicators into Statistical Programmes in Selected Latin American Countries', 5 February 2015, https://sustainabledevelopment.un.org/content/documents/13139Global%20Tracking%20Framework_World%20Bank.pdf. Also see IEA and World Bank, *Sustainable Energy for All 2015*.
[24] Our model includes the 49,053,874 persons of concern included in Tab 15 of the statistical annexes to *UNHCR Global Trends 2014: World at War*. This excludes the 5 million Palestinian refugees covered by the UNRWA.

**Figure 3: Annual energy consumption for forcibly displaced households of concern to UNHCR, 2014 (tonnes of oil equivalent)**

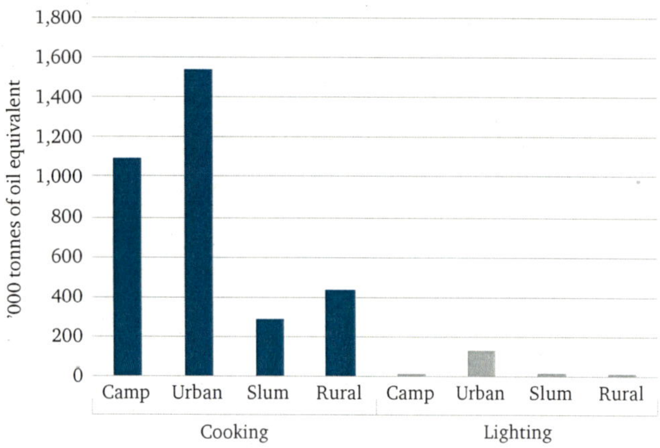


Source: Chatham House Model.

**Figure 4: Annual spending on energy for all forcibly displaced households of concern to UNHCR, 2014 ($ million per year)**

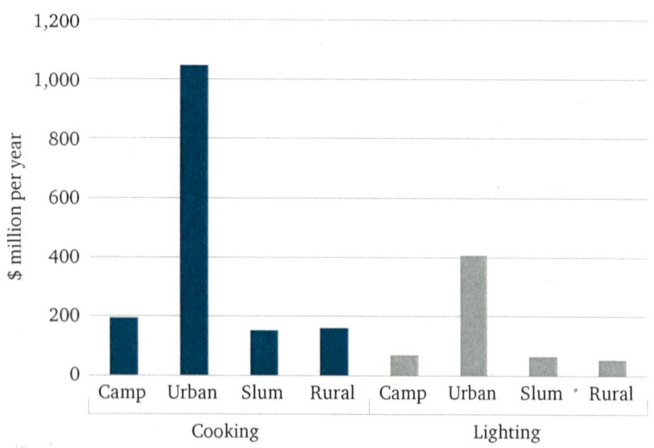


Source: Chatham House Model.
Note for both figures: For population distribution, we made a distinction between camps (which include collective centres, reception/transit camps, self-settled camps, planned/managed camps) and non-camp settings (individual accommodation and undefined/unknown). The non-camp category was then split into urban, slum and rural populations. For more information on the definition of each category, and how this split was made, see Appendices A and B.

These estimates give an idea of the scale and division of energy use among displaced households. Not all resources are paid for. Most firewood, for example, is collected and carries an invisible cost in terms of productive time lost and risks to personal safety.

Spending on cooking fuel is comparatively low in camp and rural settings. However, cooking inefficiencies mean that the volume of energy used per person is larger than the volume used by displaced people in urban areas.[25] By contrast, urban dwellers spend far more on fuel (primarily in the form of costlier LPG) in absolute terms, but use proportionately less of it because it is more efficient.

**Displaced households of concern to the UNHCR had no choice but to burn around 3.5 million tonnes of oil equivalent for cooking and lighting last year. Most of this was in the form of firewood and charcoal.**

Displaced people tend to view lighting as a secondary priority to cooking, so they use less fuel to light their homes. This is also reflected in their spending on energy. Cooking is far more fuel-intensive than lighting and is still extensively done using the 'three stone fire' method – the simplest cooking practice, whereby a cooking pot is balanced over a fire. Many refugee families cannot afford lighting. At night they live in the dark, using only the light of their cooking stoves. In the Dadaab camps in Kenya, 61 per cent of households rely on no more than a torch for lighting.[26]

Various groups pay the financial costs – including refugees and displaced people, local authorities and government, the UNHCR and responsible NGOs. In most camps and almost all non-camp situations, displaced people are responsible for meeting their own household energy needs, although assistance is sometimes extended to the most vulnerable.

A GVEP International survey in 2015 showed that 83,277 households in Kenya's Dadaab camps spent around $6.2 million in total per year on firewood. They spent $1.6 million per year on dry-cell batteries and $1.3 million per year on diesel for power.[27] The average monthly household spend on energy is $17.20.[28] These outgoings consume a significant proportion of meagre household budgets, yet the resultant energy output is inefficient. On average, individual spending on energy amounts to 24 per cent of

[25] The Chatham House model estimates that in camp settings annual consumption per person is 0.13 kilogrammes of oil equivalent (kgoe), whereas in urban settings annual consumption per displaced person is 0.05 kgoe.
[26] GVEP International field survey in Dadaab, Kenya, 2015.
[27] Ibid.
[28] Ibid.

income, compared with 55 per cent for food.[29] This provides an interesting contrast with energy spending in the host community. The average rural Kenyan household spends around 5 per cent of its income on energy and 52 per cent on food.[30] In the poorest households, even more is spent on energy. Low-income households in Nairobi's Kibera slum spend an average of $12.05 per month. Of this, 25 per cent goes on energy for lighting, 26 per cent on electricity, 22 per cent on kerosene and 27 per cent on cooking fuel.[31]

The situation is quite different in Goudoubo camp, Burkina Faso. A preliminary MEI survey discovered that its 3,053 households spend about $254,000 on firewood, $119,000 on charcoal and $16,000 on batteries per year. This equates to a monthly spend of up to $10.65 per household – 65 per cent on firewood, 30 per cent on charcoal and less than 5 per cent on batteries. Lighting expenses were lower than in Dadaab because around 60 per cent of households were using donated solar lamps.[32] Energy expenditure appears to consume between 5 and 7 per cent of individual income, compared with 40 per cent for food and 49 per cent for clothes.[33]

In several cases, agencies managing displaced populations aim to provide a set amount of cooking fuel per household. More commonly they help the most vulnerable households by reducing the need for firewood and lessening the risks to women collecting it. For example, we were told a typical three-person household among the 32,000 refugees in Burkina Faso receives 12 kilogrammes of firewood per month. However, this is not enough to cover people's needs.[34] According to the MEI survey in Goudoubo, most households use over 100 kilogrammes of firewood per month for their cooking needs.[35]

Ensuring fuel originates from sustainable sources is difficult. Sustainable fuel can be costly. For example, the Border Consortium, which manages nine camps for 120,000 Karen refugees in Thailand, spent $3.8 million in 2014 providing charcoal made from waste products to prevent deforestation. This was imported from South Korea. 'Cooking fuel is something no one wants to fund,' reported one executive.[36] This concern was regularly expressed by UNHCR camp staff we surveyed – mostly environmental officers. Figure 5 shows the Border Consortium's annual spending on charcoal. This accounts for almost 15 per cent of its total budget for 2014.[37]

**Figure 5: Border Consortium expenses for 2014 (million Thai baht)**

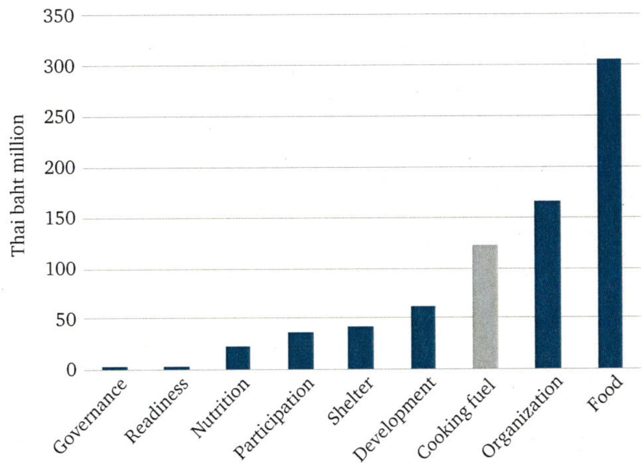


Source: *The Border Consortium Programme Report, July–December 2014.*

## Beyond financial costs

The lack of reliable energy supply takes a serious toll on health, safety and the environment. This is especially the case where refugees rely on firewood for fuel. The non-financial costs of the problem are described in more detail below. They include sexual and gender-based assaults, fire hazards in the home, electrical hazards, water contamination, illness due to inadequate heating in cold weather, indoor air pollution, poisoning, deforestation and unnecessarily high $CO_2$ emissions.

There is widespread documentation on the risk of sexual and gender-based violence faced by women and girls venturing outside camps.[38] For example, UNHCR reports show that in 63 per cent of households in Chad family members have experienced problems when collecting firewood. These problems consist of physical or verbal aggression, theft of property, rape or attempted rape,

[29] Ibid.
[30] Robert Bacon, Soma Bhattacharya and Masami Kojima, *Expenditure of Low-Income Households on Energy*, World Bank Extractive Industries for Development Series 16, June 2010, http://siteresources.worldbank.org/EXTOGMC/Resources/336929-1266963339030/eifd16_expenditure.pdf.
[31] Stephen Karakezi, John Kimani and Oscar Onguru, *Draft Report on Energy access among the Urban and Peri-Urban Poor in Kenya*, Global Network on Energy for Sustainable Development (GNESD), May 2008.
[32] Practical Action field survey in Burkina Faso, 2015.
[33] Ibid.
[34] Private correspondence with the UNHCR, Burkina Faso.
[35] Practical Action, field survey in Burkina Faso, 2015.
[36] Private correspondence with the Border Consortium.
[37] The Border Consortium, *The Border Consortium Programme Report, July–December 2014*, http://www.theborderconsortium.org/media/57485/2014-6-Mth-Rpt-Jul-Dec.pdf, p. 62.
[38] The Women's Refugee Commission (WRC) has published numerous documents on this topic. See, for example, WRC, 'Protecting Women and the Environment in the Great Lakes Region'. See also Susan C. Mapp, *Global Child Welfare and Well-Being*, Oxford University Press, 2010 , pp. 102–105.

injury or confiscation of firewood.[39] Médecins Sans Frontières (MSF) reported treating nearly 500 Darfuri women and girls in Sudan who were raped within a five-month period in 2004–05. The rapes took place during trips outside the camps to collect firewood or water.[40] Sexual violence is difficult to measure, since women are discouraged from reporting sexual assaults in many cultures and survivors fear being ostracized and punished by their communities.[41] The fact that firewood collection outside camps is illegal in many countries further encourages exploitation of the vulnerable and under-reporting of assaults.[42]

'House fires, kids' burns and hospitalization of individuals with severe burns are common, especially during the dry season when the country is dry and there are strong winds.'

Open fires, candles, illegal electricity connections and the use of kerosene for lighting all present health and safety risks. Fire can spread quickly in densely populated camps. A UNHCR camp official in South Sudan told us: 'House fires, kids' burns and hospitalization of individuals with severe burns are common, especially during the dry season when the country is dry and there are strong winds.'[43] In 2013 three separate fires in Thai refugee camps destroyed hundreds of homes, leading to a number of deaths. The causes are unknown. However, an assessment requested by the Committee for Coordination of Services to Displaced Persons in Thailand concluded: 'It was evident that causes of fires in the camps visited reflected the same trends identified elsewhere including cooking practices, candles, electrical faults or general carelessness involving the use of fire.'[44] Lack of light and power in several camp and urban situations also drives displaced people to deploy high-risk coping strategies such as power theft, with its risks of electrocution. Kerosene adds the risk of poisoning. It is often kept in plastic drinking bottles, and there are numerous cases of children drinking from them.[45]

Lack of adequate heating and insulation is a serious health risk in countries with cold winters. For example, several young children died in the freezing winter of 2014 in Lebanon and Iraq.[46] More will be affected across Europe in 2015.

Smoke inhalation in poorly ventilated cooking areas presents a further health risk to refugee and internally displaced households. The Lancet Respiratory Medicine Commission estimates that indoor air pollution in low- and middle-income countries accounts for around 3.5–4 million deaths every year.[47] Children and women have the greatest exposure to indoor pollution.[48] If we multiply our estimate of displaced people reliant on biomass (14.9 million) by the ratio of total premature deaths due to indoor pollution to the number of the world's total population relying on wood, coal or dung as their primary cooking fuel (as estimated by the WHO in 2012), indoor air pollution would be the cause of premature death for some 20,000 forcibly displaced people each year.[49] This is a broad-brush estimate subject to a number of simplifications. It does not take into account whether cooking takes place inside or outside and what other types of pollution people are exposed to. For individual camps, Chatham House used the HAPIT tool to estimate the impact of successfully substituting all traditional wood fires with ethanol stoves.[50] Particulates would be reduced

[39] GVEP International Chad assessment, 2015.

[40] MSF, 'Rape and sexual violence ongoing in Darfur, Sudan', 3 July 2005, http://www.msf.org/article/rape-and-sexual-violence-ongoing-darfur-sudan.

[41] Mapp, *Global Child Welfare and Well-Being*, pp. 102–105.

[42] Private correspondence with Global Alliance for Clean Cookstoves (GACC).

[43] Private correspondence with UNHCR, South Sudan.

[44] Operation Florian, 'Operation Florian undertake fire assessment of refugee camps in Thailand', 11 January 2015, http://www.operationflorian.com/operation-florian-undertake-fire-assessment-of-refugee-camps-in-thailand/.

[45] Comparing the different estimates, we see that between 50 and 75 per cent of all child poisoning in off-grid areas is caused by kerosene ingestion. David C. Schwebel et al., 'Paraffin-related injury in low-income South African communities: knowledge, practice and perceived risk, *Bulletin of the World Health Organization,* September 2009, Vol. 87, No. 9, pp. 700–706, doi: 10.2471/BLT.08.057505. Also see Evan Mills, 'Health Impacts of Fuel-based Lighting', Lumina Project Technical Report #10, 16 October 2012, http://light.lbl.gov/pubs/tr/lumina-tr10-summary.html.

[46] IFRC, 'Syrian refugees among eight dead as ferocious snow storm hits Lebanon', 27 January 2015, http://www.ifrc.org/en/news-and-media/news-stories/middle-east-and-north-africa/lebanon/syrian-refugees-among-eight-dead-as-ferocious-snow-storm-hits-lebanon-68004/; Christian Aid Mission, 'Winter Brings Killer Cold to Those Who Fled Death in Middle East', 15 January 2015, http://www.christianaid.org/News/2015/mir20150115.aspx.

[47] Stephen B. Gordon et al., 'Respiratory risks from household air pollution in low and middle income countries', *The Lancet Respiratory Medicine*. Vol. 2, No. 10, October 2014, http://dx.doi.org/10.1016/S2213-2600(14)70168-7, pp. 823–60.

[48] For example, almost 50 per cent of pneumonia deaths among children under five are found to be due to particulate matter inhaled from indoor air pollution. See Strohmeier, *Why sustainable energy matters to children*.

[49] This back-of-the-envelope calculation first took the ratio of deaths as a result of indoor air pollution – 4.3 million people annually as estimated by the WHO – to the total number of people dependent on solid biomass globally – 2.9 billion as estimated by the World Bank. This ratio was then applied to the number of displaced people we estimate to be reliant on solid biomass. Sources: WHO, 'Household air pollution and health', Fact sheet No. 292, updated March 2014, http://www.who.int/mediacentre/factsheets/fs292/en/; World Bank, 'Unlocking Clean Cooking and Heating Solutions Key to Reaching Sustainable Energy Goals', feature story, 19 May 2015, http://www.worldbank.org/en/news/feature/2015/05/19/unlocking-clean-cooking-and-heating-solutions-key-to-reaching-sustainable-energy-goals.

[50] Household Air Pollution Intervention Tool 2.0, GACC and Household Energy, Climate, and Health Research Group at University of California, Berkeley, https://hapit.shinyapps.io/HAPIT.

by 85 per cent as a result.[51] Our findings suggest such an intervention would save around 86 lives in the Dadaab camps in the first year and another 200 lives thereafter.[52]

Energy use patterns among displaced people create considerable environmental costs. Figure 6 shows the breakdown of $CO_2$ emissions in terms of refugee cooking and lighting globally.

**Figure 6: Annual $CO_2$ emissions produced by all forcibly displaced households of concern to the UNHCR, 2014[53] (million $tCO_2$)**

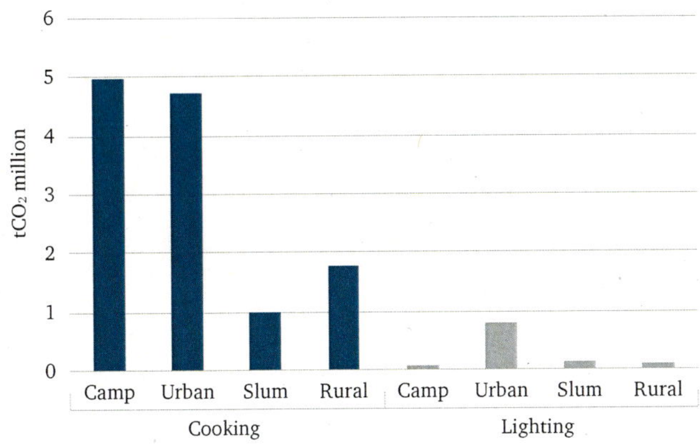


Source: Chatham House Model.
Note: See Appendix B for 'urban', 'rural' and 'slum' definitions.

Deforestation is a major problem in many countries hosting large numbers of forcibly displaced people. For example, the Democratic Republic of the Congo has one of the world's largest expanses of tropical forest but an accelerating rate of deforestation.[54] The country hosts 4 million people of concern to the UNHCR.[55] This accounts for nearly 7 per cent of the total number of forcibly displaced people worldwide.

Our study estimates that an area of forest equivalent to 49,000 football pitches is used each year to produce energy just for refugee camp populations.[56] This number would increase considerably if it factored in non-camp populations, which are much larger and whose energy use is more difficult to assess.

Most refugee impact is minor compared with the impact of illegal or unsustainable logging in host countries. Refugees themselves rarely cut down trees unless employed by loggers. They chiefly rely on deadwood and branch collection. However, where this is the main source of fuel, demand for locally produced firewood and charcoal is likely to grow, and perceptions of competition for resources can cause resentment.

According to the regional director of the Ministry of Environment in Dori, Burkina Faso, growing populations and an influx of refugees have played a role in increasing desertification. More wood is being harvested on a small scale without a permit. The natural barrier of wooded land between the Sahel and the Dori flood plain is becoming ever thinner due to increasing numbers of grazing livestock owned by nomadic refugees. To counter this trend, the wood distributed to camp populations is now collected from an area 100 kilometres south of Dori and then transported to the camps.[57]

## Camp administration and operations

Humanitarian agencies, local/national governments and NGOs spend money on a range of items and operations that require energy to deliver essential services to refugee populations. These include infrastructure equipment such as water pumps, filters and streetlights, and facilities such as schools, community centres, hospitals and camp offices. Fuel is also required to transport camp staff (who rarely live inside the camps), food, fuel, shelter, medical equipment and sometimes fresh water.

Data on energy costs and usage for administration and operations in the humanitarian sector are not pooled or collected in any standardized way. The majority of operations rely on diesel generation. Fuel is likely to be paid for by a number of different organizations and may be allocated different budget lines within the same organizations. In the case of the UNHCR, fuel for

[51] American Public Health Association, 142nd Annual Meeting & Expo, 17 November 2014, 'Addressing Refugee Health and Safety through Gender-Specific Interventions: Clean Ethanol Stoves Fuel as a Tool for Protection and Prevention', https://apha.confex.com/apha/142am/webprogram/Paper303850.html, last accessed on 20 October 2015.
[52] Household Air Pollution Intervention Tool 2.0.
[53] $CO_2$ conversion factors for cooking (unit used = kg $CO_2$/kg primary fuel): charcoal = 1.8, firewood = 1.8, processed solid fuel = 1.4, biogas = 1.2, LPG = 2.9, kerosene = 3.0. The $CO_2$ emission factors used for cooking are gross emissions at point of combustion. They do not take into account the extent to which sources of biomass for fire might be sustainable. $CO_2$ conversion factors for lighting (unit used = kg $CO_2$ per kilowatt-hour unless otherwise specified): torches = 0, kerosene = 2.96 kg $CO_2$ per litre, electricity = country-specific on-grid electricity carbon emissions (GTF data), mini-grid 1 = 0.36, mini-grid 2 = 1.21, diesel generator = 1.21, solar = 0. For further details on the emissions factors used, please contact the authors.
[54] Sam Lawson, *Illegal Logging in the Democratic Republic of the Congo*, April 2014, Chatham House, http://www.chathamhouse.org/sites/files/chathamhouse/home/chatham/public_html/sites/default/files/20140400LoggingDRCLawson.pdf.
[55] UNHCR, '2015 UNHCR country operations profile – Democratic Republic of the Congo', http://www.unhcr.org/pages/49e45c366.html, last accessed on 20 October 2015.
[56] Based on the interviews conducted by Chatham House and field research conducted by partners, we derived an average household firewood consumption figure for firewood-dependent households (119.16 kg per household per month). Chatham House scaled this up to the total population of concern to the UNHCR living in camps assigned to Cooking Type 1 (firewood-dependent), as this cohort (numbering 6,644,004 people) would by definition use the most firewood for cooking. These assumptions give total firewood consumption of 1.9 million tonnes/year. This figure was then converted via cords and acres to 'football fields' of forest. For further details on this calculation and the model, please contact the authors.
[57] Practical Action field survey in Burkina Faso, 2015.

powering offices and compounds comes from the UNHCR mission's administration budget. The energy costs of other camp facilities (schools, hospitals, refugee community facilities, etc.) and infrastructure items (water pumps, filters, streetlights, etc.) are charged to the UNHCR mission's operations budget. Many services are contracted out to implementing partners. It is not clear to what extent implementing partners are required to submit breakdowns of their fuel and energy costs to the UNHCR. There are also other local authorities, government agencies and NGOs that provide services to displaced people. Each agency will have its own method of accounting for and reporting energy use and spending.

'If generator systems were better understood and systems developed around their upkeep, the device would be more efficient, consume less fuel and need less spare parts.'

Anecdotally, it seems that the way that electrical services are dealt with in most NGOs, implementing partners and the UN is inadequate and that there is a shortage of skilled technicians. Basic servicing of generators is often outsourced without any major tracking systems or checklists in place, and little quality control is deployed in the selection of contractors to carry out electrical work. Accounts from the field suggest that generators are often over- or underloaded, which can reduce efficiency.[58] One electricity specialist working in the field said, 'If generator systems were better understood and systems developed around their upkeep, the device would be more efficient, consume less fuel and need less spare parts. The same goes for correct wiring systems, better wiring and upkeep means no phase balancing issues and better disconnections time (safer systems) plus less heat/ expansion and contraction issues.'[59]

Interviewees frequently expressed a desire to reduce diesel costs. They were interested in either connecting to the national electricity grid or introducing or scaling up solar equipment or solar/diesel hybrid systems for camp operations. However, a prerequisite for designing more cost-effective energy solutions for camp administration and operations is to get an accurate picture of current diesel costs across different camps and locations. This remains difficult. For example, when interviewees were asked to estimate the fuel bills for their operations, the following issues arose:

- Respondents generally excluded diesel use (or costs incurred) by implementing partners. Given that implementing partners often have significant involvement in camp operations – providing services such as water pumping – this is a major omission.
- In some cases, diesel volumes were known but were simply multiplied by the local diesel price rather than by the total cost of that supply to the camp authorities. The latter would include the cost of transportation and a margin for the supplier.
- In other instances, the number cited was a cost paid to a company or partner for the diesel supply. This would include its transportation and perhaps generator maintenance.

Costs for delivery of fuel, food and other supplies along with personal transportation were either excluded altogether or not separated from the figures provided.

In short, the fact that detailed data on diesel purchase, distribution and use are not available or are integrated into other figures demonstrates the lack of awareness of the value of energy accounting throughout the humanitarian system. Due to this data limitation, we can make only general points about energy use in the operations of humanitarian agencies.

As mentioned, electricity for humanitarian agencies' compounds generally comes from diesel generators. This method of energy provision carries high associated costs and there are wide disparities in use. To cite two small examples, the UNHCR bought 85,962 litres of diesel for the camps across Ethiopia (729,460 refugees)[60] between 2013 and the first quarter of 2015.[61] Its implementing partners will have spent more. Estimates of total use at Azraq camp in Jordan (about 26,000 refugees) are in the region of 2 million litres per year, the main consumers being the hospital and the supermarket. We were able to retrieve more detailed data on energy costs for the operation and administration of Goudoubo refugee camp in Burkina Faso and the Dadaab refugee camps in Kenya. See Figures 7 and 8 below.

[58] Evidence from field visits during this project also suggests that too many diesel generators may be operating within refugee camps. Some redundancy in the system is obviously useful. However, systematically underloading generators not only incurs excess capital expenditure on the equipment, but also leads to associated operating and maintenance problems. Based on field surveys by GVEP International and Practical Action, and private correspondence with UNHCR camp managers.
[59] Private correspondence with UNHCR HQ.
[60] UNHCR field staff interview and verification with UNHCR headquarters. See UNHCR, '2015 UNHCR country operations profile – Ethiopia', http://www.unhcr.org/pages/49e483986.html.
[61] The period was 15 months in total, according to an interview with a UNHCR camp manager in Ethiopia.

**Figure 7: Annual spending on camp administration and operations compared to refugee household spending on energy (cooking and lighting) at the Dadaab refugee camps, Kenya**

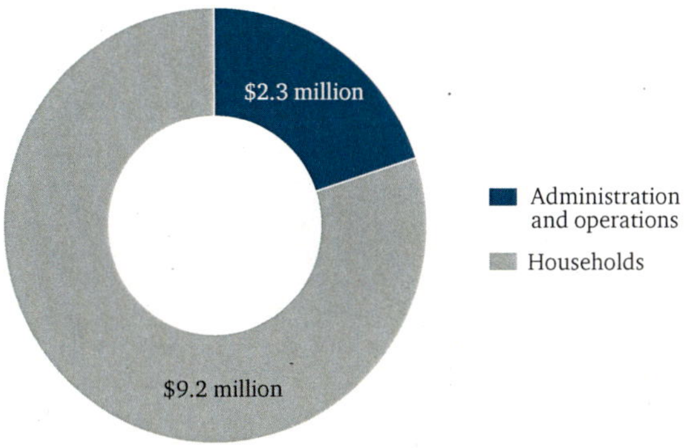


Source: Interviews with camp operators and refugee households at the Dadaab refugee camps, by GVEP International for the MEI, 2015.

**Figure 8: Annual spending on camp administration and operations compared to refugee household spending on energy (cooking and lighting) at Goudoubo refugee camp, Burkina Faso**

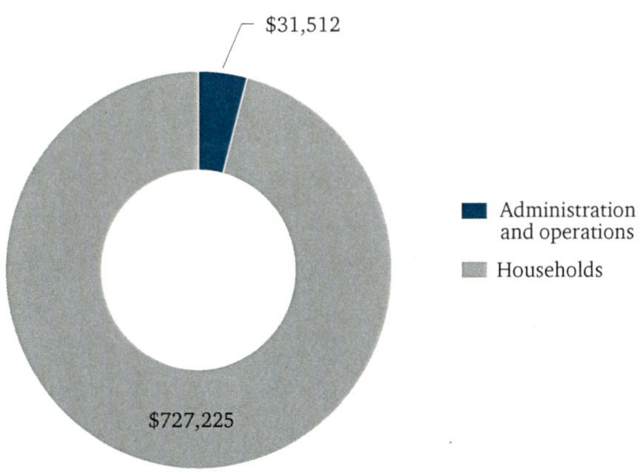


Source: Interviews with camp operators and refugee households at Goudoubo refugee camp, by Practical Action for the MEI, 2015.

The figures indicate that spending on camp infrastructure and administration in Dadaab accounts for approximately 20 per cent of spending on energy in the whole camp every year. In Goudoubo this figure is much lower, representing only 4 per cent of spending on energy in the camp.

If every camp followed Dadaab's pattern of diesel consumption per camp resident, around 59 million litres would be consumed globally each year to run refugee camps. However, if every camp followed the Goudoubo pattern, the corresponding figure would be 18.8 million litres. Interviews conducted during this project indicate spending levels at Goudoubo are more normal in sub-Saharan Africa and parts of Asia, where energy infrastructure is usually minimal.

Both Dadaab and Goudoubo have some way to go to reach appropriate levels of access. Indeed, the lack of electricity seriously inhibits the welfare of inhabitants. No electricity is provided in Goudoubo except to power the pumping stations and health centre and light the school (all supplied by diesel generators). On-site administrative offices are without power. As a result, implementing partners such as International Emergency and Development Aid (IEDA Relief), which is responsible for camp management, lack computers and even lights for their offices. Most activities have to be conducted using pen and paper or mobile phone. Public lighting is not available to the displaced community. Economic, educational, social and other activities are thus limited to daylight hours. As a point of comparison, the rural electrification rate in Burkina Faso is only 1.4 per cent.[62] Any improvements to camp electrification should, therefore, be extended to the local population.

Where camps are connected to the electricity grid, costs per kilowatt-hour (kWh) may be lower than those of diesel generation. Nevertheless, total costs may well be higher as more plentiful energy supply prompts higher consumption. A handful of camps are connected to the national electricity grid, which provides operators with a more efficient, stable and cost-effective source of energy. These are mainly in western Asia and are situated close to urban centres. However, costs can spiral if households and businesses connect informally to the grid without cost recovery mechanisms being in place, as was the case in Zaatari, a refugee camp in Jordan hosting Syrians displaced by the war (see Box 4).

While transportation is not the focus of this report, there may be considerable scope for lowering these costs too. Since many of the camps are in remote locations, fuel transportation adds to energy costs. As a point of reference, a study for the US army estimated that for every litre of fuel used in remote bases, six litres were consumed transporting it to the bases.[63] Transport and contracting fees often increase the costs to humanitarian

[62] World Bank and International Finance Corporation, 'Lighting Africa – Burkina Faso', https://www.lightingafrica.org/where-we-work/burkina-faso/.
[63] Army Environmental Policy Institute, *Sustain the Mission Project: Resource Costing and Cost-Benefit Analysis* (Arlington, VA: Army Environmental Policy Institute, 2006).

agencies. More effective energy solutions in camps and remote locations could therefore lessen the need for diesel trucking. Increased use of local materials and food produce may also merit inclusion in any strategy to reduce transport costs. In the future, technological innovations such as 3-D printing may offer opportunities to produce shelter and other necessary items on site.[64] In many countries hosting displaced populations, fuel is subsidized by the government. Host governments, which may be footing some of the transport bill, could therefore also benefit financially from improved systems. However, reforming these systems will be a complicated task, given the variety of groups that may be involved in marketing and delivering fuel, their contractual costs and the monopolies that form around these activities. In-depth research with a focus on transport in the humanitarian sector is needed to fully capture the costs involved and assess the potential for change.

### Box 4: Two examples of high spending on camp operations

Zaatari camp in Jordan is a somewhat exceptional case in terms of energy access. This is because operations at the camp – which housed around 84,000 Syrian refugees in 2014 – were connected to the national electricity grid when it was set up in July 2012. The refugees, who were accustomed to having electricity in Syria, began to tap into the grid informally through the camp's street lights to power their homes and businesses. The informal connections increased consumption, while the regional power distributor charged the UNHCR at the full commercial rate. Zaatari's 2014–15 electricity bill came to $8.7 million. The unsustainable costs led the agency to cut off informally connected households and businesses in 2015. This resulted in some businesses moving to the use of diesel generators. A new system is being put in place with metering to ensure each household has access to a basic amount of energy. Income-earning businesses can pay for additional energy.[a]

Camps in the South Sudan counties of Maban and Pariang house approximately 226,000 refugees. The camps incur considerable diesel costs because they lack grid connections and have a large number of refugees. South Sudan's public infrastructure is poor. The whole country is 620,000 square kilometres in area – twice the size of Burundi, Uganda and Rwanda put together – but it has only 250 kilometres of paved road.[b] Most of these roads are concentrated around the capital, Juba. The Sudd, a vast and unique area of swampland, is a natural barrier between the North and the South and makes land travel nearly impossible in the rainy season. The diesel used in Maban and Pariang is refined in Mombasa and usually transported some 2,800 kilometres by lorry to Juba. When the Sudd floods, the fuel is taken by plane from Juba to the camps. The MEI can only speculate about the final transportation costs. However, it is clear that these will be extraordinarily high, and certainly significantly higher than the costs of fuel in camps themselves.[c]

[a] Private correspondence and interviews with the UNHCR, Jordan.
[b] African Development Bank Group, *South Sudan: An Infrastructure Action Plan – A Program for Sustained Strong Economic Growth*, 'Chapter 7: Transport Services and Infrastructure', p. 170.
[c] It should be highlighted that all UN agencies and other humanitarian actors working in these areas of South Sudan face tremendous logistical difficulties. High transportation costs will be borne by all groups including the WFP, the UN Mission in Sudan (UNMIS) and others.

[64] See, for example, Imogen Mathers, '3D printing can revolutionise emergency healthcare', SciDev.Net, 17 August 2015, http://www.scidev.net/global/design/multimedia/3d-printing-emergency-healthcare-haiti-maternity.html.

# 3. What Do the Numbers Tell Us?

The findings on current usage, lack of access, costs and human and environmental impacts described in the last chapter indicate that there is an opportunity to reap multiple benefits from managing energy differently. This chapter examines the business case for shifting the current inefficient biomass- and fossil fuel-based systems towards cleaner and more sustainable ones while improving levels of access. Using the model built by Chatham House, the latter part of this chapter makes a broad-brush assessment of costs, savings and trade-offs based on three scenarios for change in energy use in displaced households.

## Human protection, health and environmental multipliers

Simple energy interventions have the capacity to deliver a range of benefits for displaced populations. Focusing on energy access gives the humanitarian sector the opportunity to improve a range of outcomes for forcibly displaced people while simultaneously supporting host governments' sustainability goals. These are summarized in Figure 9.

Many of the benefits listed in Figure 9 are explained in more detail elsewhere.[65] Here we give just some examples.

Clean cookstoves have produced demonstrable reductions in firewood use by refugee households, normally amounting to 30–70 per cent.[66] There are also a known range of health benefits, particularly in terms of reducing acute respiratory infections (ARIs).[67] Cookstove projects often concentrate on protecting children and women, who are almost always responsible for collecting firewood and who often travel long distances in unsafe surroundings to do so.[68] A survey of a three-month pilot study in an Ethiopian refugee camp found that introducing efficient cookstoves in 100 households reduced the total time spent collecting firewood from 1,659 hours to 732 hours per month.[69]

[65] See Gunning, *The Current State of Sustainable Energy Provision for Displaced Populations*, for a good overview.

[66] See, for example, Ahmed Hassan Hood, 'Evaluation Report: Efficient Stoves for IDPs Protection and Environment Conservation', Sudanese Agency for Environment and Development (SAEDS), 2007.

[67] See, for example, http://www.who.int/indoorair/health_impacts/disease/en/; and Christopher Rogers, Benjamin Sovacool and Shannon Clarke, 'Sweet nectar of the Gaia: Lessons from Ethiopia's "Project Gaia"', *Energy for Sustainable Development*, 2013, Vol. 17, No. 3.

[68] Gunning, *The Current State of Sustainable Energy Provision for Displaced Populations*, p. 31.

[69] Amare Egziabher, James Murren and Cheryl O'Brien, *An Ethanol-fueled Household Energy Initiative in the Shimelba Refugee Camp, Tigray, Ethiopia: A Joint Study by the UNHCR and the Gaia Association, UNHCR and Gaia Association*, 20 January 2006, https://www.projectgaia.com/files/ShimelbaCampGAUNHCR.pdf.

## Cut costs

Widespread introduction of improved cookstoves and basic solar lanterns could save $323 million a year in fuel costs.

## Reduce emissions

Widespread introduction of improved cookstoves and basic solar lanterns would reduce emissions by an estimated 6.85 million $tCO_2$ a year.

**Figure 9: Benefits from sustainable energy provision to displaced people**

| | | | | | |
|---|---|---|---|---|---|
| **Environmental benefits** | *Avoidance of deforestation and environmental degradation* | *Preservation of biodiversity* | *Reduced $CO_2$ emissions* | *Reduced risk of contamination and local pollution* | |
| **Security and protection** | *Lower risk of gender-based violence outside camp* | *Reduction of night-time violence in camp* | *Lower risk of fire-related accidents* | | |
| **Health** | *Reduction in indoor pollution* | *Reduction in risk of burns* | *Improved nutrition* | *Improved availability of clinical services* | |
| **Livelihoods and resilience** | *Time saved and longer hours available for other activities* | *Less reliance on diminishing local resources* | *Diversification of activities* | *Money saved* | *Improved education opportunities* |
| **Benefits for host populations** | *Less competition for similar services* | *Reduction in reliance on limited natural resources* | *Improved livelihoods* | | |
| **Economic and energy security benefits** | *Cost savings* | *Improved energy security* | *Better reliability* | *Reduced fuel price exposure* | *Increased operational lifetime* |

Refugees have consistently called for improved lighting to make them feel safer[70] and increase their ability to read and study at night.[71] In Goudoubo refugee camp in Burkina Faso, 53 per cent of respondents to the MEI's household survey said public street lighting would be 'very important' and 43 per cent said it would be 'important'. The same survey revealed that in only 4 per cent of households would females go out after dark. In 18 per cent of households no one would leave the tent after dark.[72]

Refugees have consistently called for improved lighting to make them feel safer and increase their ability to read and study at night.

Camp managers interviewed by the authors consistently emphasized the financial savings they felt could be made by using sustainable energy. Several camp managers pointed to the large savings likely to be achieved by pumping water using solar hybrid systems rather than diesel alone. The UNHCR has conducted a number of assessments on switching to solar water pumps in East Africa and the Horn of Africa. Innovation Norway has also recently made donations to support a hybrid system for water pumping in camps in Maban, South Sudan.[73] Commitments have also been made to evaluate and pilot hybrid technologies for office and staff accommodation and to promote changes in staff behaviour. For example, the UNHCR has stopped paying for individual staff transportation to and from camps in Jordan and is encouraging mass transport solutions.

Measures to reduce fuel use and inefficiencies in camp administration and logistics are suggested in Table 2 below.

## Three scenarios: additional capital costs and annual cost savings

So what kind of costs and savings are we talking about in making changes to household energy delivery? The model developed by Chatham House demonstrates a clear financial incentive for sustainable energy interventions by estimating the global effects of particular interventions and the broad financial and emissions costs and savings associated with them. However, data are limited, and a number of caveats must be noted (see Appendix A for the methodology behind the model).

This model does not advocate one particular scenario over any other. Indeed, energy solutions are specific to each context and need to be evaluated on a case-by-case basis. For example, in circumstances where LPG is produced locally, there may be a stronger case for introducing it. Costs will be much higher in places were LPG has to be imported and no local supply chains or knowledge exist.

### Table 2: Examples of sustainable energy solutions to meet camp operation energy needs

| Energy services | Energy efficiency options | Renewable energy options | | Other options/low carbon | |
|---|---|---|---|---|---|
| Health, education and community centres | • Improved institutional cookstoves (ICS)/Fuel-efficient stoves (FES)<br>• Insulated buildings<br>• Energy management | • Alternative fuel and stoves – biomass briquettes, solar, ethanol, biogas<br>• Stand-alone PV systems<br>• Solar AC/chillers/solar water heating | • Mini- and micro-grid electricity from wind, biomass, biogas, PV, hybrids | • Alternative fuel and stoves – LPG<br>• Energy storage | |
| Security, lighting, communication, water pumping and waste management | • Energy management<br>• Increased efficiency | • PV street lighting<br>• Biogas systems<br>• PV water pumping<br>• Stand-alone PV systems | • Mini- and micro-grid electricity from wind, biomass, biogas, PV, hybrids | • Mini-grid electricity from more efficient gensets<br>• Energy storage<br>• Grid electricity | |
| Administration and logistics | • Energy management<br>• Increased efficiency<br>• Locating staff accommodation closer to site<br>• Encouraging group transport options to and from sites | • Solar water heating<br>• Biogas systems<br>• Stand-alone PV systems | • Mini- and micro-grid electricity from wind, biomass, biogas, PV, hybrids | • Alternative fuel and stoves – LPG | • Grid electricity<br>• Mini-grid electricity from more efficient gensets |

Source: Gunning, *The Current State of Sustainable Energy Provision for Displaced Populations*, p. 44.

[70] UNHCR, *Light Years Ahead: Innovative Technology for Better Refugee Protection*, March 2012.
[71] Ibid.
[72] Practical Action, field survey in Burkina Faso, 2015.
[73] Private correspondence with Paul McCallion, UNHCR.

Three broad-brush scenarios of rising levels of ambition are outlined below: **Incremental Change, Alternative Energy** and **Fundamental Change.** These are based on step changes in tier energy access summarized in Table 3 (reproduced from the World Bank and ESMAP's Energy Sector Management Assistance Program).

## The Incremental Change scenario

The Incremental Change scenario describes a situation in which all forcibly displaced people adopt more efficient cooking equipment and have access to simple lighting solutions. Since a large number of displaced people are dependent on firewood, the scenario broadly describes the widespread introduction of improved cookstoves for cooking (minimum Tier 3 access). It also describes a combination of basic solar lanterns and diesel to meet lighting needs (minimum Tier 1 access). As with all the scenarios, the UNHCR and/or any relevant private–public partners would pay the upfront capital costs. The capital costs here only include the costs of providing the technology and exclude the crucial costs of maintenance and supply. The year-on-year savings generated would accrue mainly to forcibly displaced households.

Figure 10 below reveals that under the Incremental Change scenario displaced people could save $323 million each year after an initial capital investment of $335 million from humanitarian agencies.[74] When considered alongside the potential health, livelihood and protection benefits, annual cost savings as high as this reveal the huge opportunity to the international community.

### Table 3: Simplified tiered matrix of energy access

| Attributes of energy supply | | Tier 0 | Tier 1 | Tier 2 | Tier 3 | Tier 4 | Tier 5 |
|---|---|---|---|---|---|---|---|
| Capacity | Household electricity | No electricity[a] | Very low power | Low power | Medium power | High power | |
| | Household cooking | Inadequate capacity of primary cooking solution | | | | Adequate capacity of primary cooking solution | |
| Duration and availability | Household electricity | <4 hours | 4–8 hours | | 8–16 hours | 16–22 hours | >22 hours |
| | Household cooking | Inadequate availability of primary cooking solution | | | | Adequate availability of primary cooking solution | |
| Reliability | Household electricity | Unreliable energy supply | | | | Reliable energy supply | |
| Quality | Household electricity/ cooking | Poor-quality energy supply | | | Good-quality energy supply | | |
| Affordability | Household electricity | Unaffordable energy supply | | Affordable energy supply | | | |
| | Household cooking | Unaffordable energy supply | | | | Affordable energy supply | |
| Legality | Household electricity | Illegal energy supply | | | Legal energy supply | | |
| Convenience | Household cooking | Time and effort spent sourcing energy cause inconvenience | | | Time and effort spent sourcing energy do not cause inconvenience | | |
| Health and safety | Household electricity | Unhealthy and unsafe energy system | | | | Healthy and safe energy system | |
| | Household cooking[b] | Level 0 | Level 1 | Level 2 | Level 3 | Level 4 | Level 5 |

[a] The detailed multi-tier matrix for household electricity considers a continuous variable between Tier 0 and Tier 1 for basic lighting services so as to capture the contribution of solar lamps that do not reach the minimum output threshold required for Tier 1 access but that are highly affordable and enable households to reduce or eliminate the use of kerosene for lighting.

[b] Levels are defined based on the technical performance of the cookstove (for example, in terms of efficiency, pollution and safety), kitchen ventilation and conformity of usage (use of required accessories, regular cleaning and so on.)

Source: Energy Sector Management Assistance Program, http://www-wds.worldbank.org/external/default/WDSContentServer/WDSP/IB/2015/02/27/090224b082b6d2b4/2_0/Rendered/PDF/Capturing0the00ity0of0energy0access.pdf.

[74] The initial capital cost is calculated on the basis of the technology costs of providing improved cookstoves. As mentioned above, it does not include distribution and training costs etc. For cooking, the savings that accrue to the displaced populations assume they maintain the same level of consumption. Increased levels of efficiency then reduce the amount of fuel required. The model assumes a fixed cost for firewood that is often not financial in nature. Firewood is often free. The lighting savings are lower but primarily arise from replacing batteries and kerosene with solar lanterns, eliminating the cost of usage altogether. More details are included in the methodology in Appendix A.

However, the Incremental Change scenario indicates only a minimum level of ambition. Indeed, the toolkit on cooking technologies that accompanies this report highlights WHO guidelines on household fuel which suggest that a move away from solid biomass for cooking is the only path towards globally acceptable standards of safety.[75]

**Figure 10: Potential savings and capital cost – widespread introduction of clean cookstoves and basic solar lanterns**

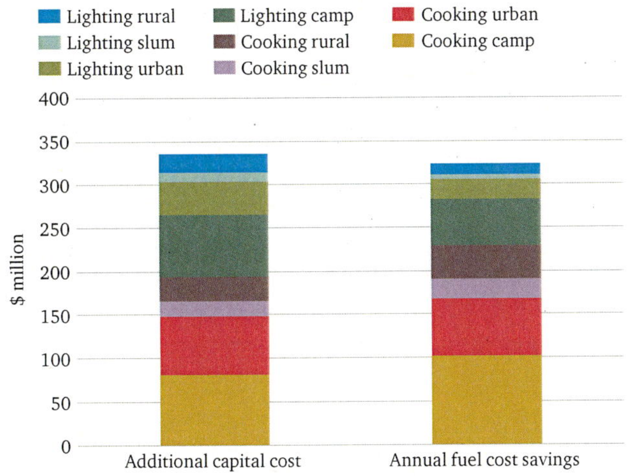


Source: Chatham House Model.

## The Alternative Energy scenario

The Alternative Energy scenario describes the widespread introduction of biomass briquettes to provide a minimum of Tier 3 access for cooking and a combination of solar lanterns and mini-grid lighting solutions. This scenario would cost more to implement globally than the Incremental Change scenario. However, it would offer significant benefits on the previous scenario – albeit while failing to eliminate some concerns and adding a few more.

While the Incremental Change scenario outlines only minimal targets for improving the sustainability of lighting solutions, the Alternative Energy scenario is more ambitious. It provides all displaced people with sustainable lighting, with many connected to mini-grids. However, access remains limited to Tier 1, amounting to only a few hours of lighting per day. The cooking scenario involves the widespread introduction of biomass briquettes, which would ease resource pressure on firewood consumption in and around refugee camps. Biomass briquettes are also often used in conjunction with associated work programmes and can thus contribute to livelihood activities and the generation of local markets. Despite this, biomass briquettes fail to meet WHO safety guidelines on indoor air pollution. Many of the problems that relate to the introduction of improved cookstoves thus remain relevant to this scenario. Experience also suggests that processes for manufacturing briquettes can be very poor if not well planned and managed. This often defeats the purpose of the intervention, since high moisture content and/or poor gasification can diminish efficiency savings.[76]

Figure 11 shows that the upfront capital costs of providing the technology alone for this scenario would be $769 million. Breaking this down further reveals that the costs of cooking solutions ($194 million) would be comparable to those in the Incremental Change scenario, but that the costs of the more advanced lighting solutions would be higher ($574 million).

The lighting solution outlined in this scenario would cost more but would also generate annual savings of $117 million for forcibly displaced people. By contrast, the model suggests that annual cooking fuel costs would increase by $70 million per year for displaced households if they paid the full market price. This indicates that biomass briquette programmes of this kind may need to be implemented with close attention to fuel affordability and perhaps subsidies and market development strategies. This has been a typical feature of biomass briquette pilot projects.[77]

**Figure 11: Potential capital cost and savings – widespread introduction of biomass briquettes and solar and mini-grid lighting solutions**

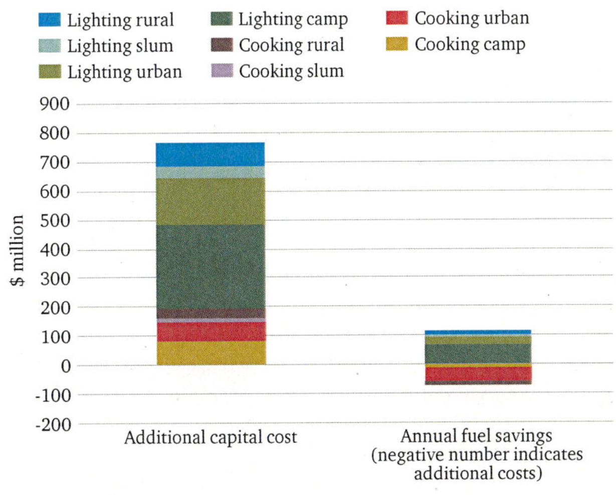


Source: Chatham House Model.

[75] For more on this see Practical Action, 'Toolkit on cooking technologies', 2015. Available at: https://www.chathamhouse.org/about/structure/eer-department/moving-energy-initiative-project.
[76] Private correspondence with Paul McCallion, UNHCR.
[77] See the example of Nakivale refugee camp in Uganda outlined briefly in UNHCR, 'Innovation: Briquette-making project helps protect women in Ugandan camp', 9 August 2013, http://www.unhcr.org/520500559.html; or the example of compressed rice husk distribution in Kutupalong, Bangladesh, UNHCR, 'World Environment Day: Putting refugee protection first pays dividends for the environment in Bangladesh', 4 June 2015, http://www.unhcr.org/4c08eacb6.html.

## The Fundamental Change scenario

The Fundamental Change scenario aims for the widespread introduction of LPG to provide Tier 3 access for cooking and a more ambitious tier level target for lighting than under the Alternative Energy scenario. Lighting in the Fundamental Change scenario again consists of a combination of solar and mini-grid solutions, as with Alternative Energy, but provides a higher tier of access. Solutions included here represent a fundamental upgrade to the modern energy supply that must be considered the ultimate objective of humanitarian agencies. However, it will take time and financial support to deliver them successfully.

Indeed, the widespread introduction of LPG for forcibly displaced people is highly ambitious. It is not always practical since the fuel itself is more expensive, and consolidating supply chains and infrastructure is both time-consuming and costly. Figure 12 shows that initial capital costs for LPG cookstoves are estimated at $319 million. This excludes other infrastructure or supply chain costs. The annual cost of fuel would add around $786 million per year. Again this shows that some subsidies and market development strategies would be essential for the introduction of LPG solutions. Nonetheless, LPG is a highly efficient cooking alternative to firewood, charcoal and biomass briquettes. It conforms to WHO standards on safety, and thus also offers associated benefits for health and protection. The support required will vary widely and for some producers of LPG – especially where deforestation is driving up firewood costs – the business case will be clearer. Box 5 gives an example of an LPG project in Sudan where local fuel costs for charcoal and wood were two and three times higher than the estimates in our model, while the monthly household LPG cost was just half.

### Box 5: The cost-effectiveness of replacing wood fuels with LPG using carbon financing in North Darfur

For the past six years Practical Action, an NGO, has been working to facilitate the development of a local liquefied petroleum gas (LPG) market in El Fasher in North Darfur. The area includes peri-urban and rural villages and IDP camps. One of the main objectives of the Darfur Low Smoke Stoves project is to combat deforestation in the area. Forest resources have been rapidly depleted over the last 30 years, exacerbating conflict over resources,[a] and harvested fuelwood and charcoal remain the main sources of household energy for over 90 per cent of the population.[b] Increasing scarcity and conflict in the area mean that people are less able to collect firewood and must rely on local markets for cooking fuel. The firewood on sale there is transported from further and further away (now up to 180 kilometres in 2015), adding to the final price paid by households. The price of firewood per kilogramme more than doubled from 0.47 Sudanese pounds (SDG) to 1.25 SDG ($0.28) between 2010 and 2013.[c] This is over three times the price in our model. In 2013 charcoal traded at 1.71 SDG/kg ($0.38/kg) – more than four times its 2010 price and over double the price in our model.[d] In 2013 a 70-kg sack of charcoal, which lasts the average household approximately 30 days, cost around 120 SDG ($26.98). This translates into an annual average household fuel cost of about $324.

In the context of rising biomass costs and domestic LPG production, Practical Action found that LPG could cost-effectively replace wood and charcoal. The initial capital cost of a 12.5-kg LPG canister and stove is 700 SDG ($157). The monthly fuel cost comes to 50 SDG ($11.24), half of the monthly fuel cost for LPG included in our model.[e] In the first year, therefore, households would spend $292. In subsequent years they would spend $135. Based on 2013 prices, if households switch entirely from charcoal to LPG they save about $32 in the first year and $189 in the years following. The savings from switching, therefore, cover the initial capital cost of the stove and canister in less than two years.

The Low Smoke Stoves Project was started with an investment of about £250,000 from CarbonClear Ltd, a carbon management firm. The money was used to set up a revolving fund managed by a local NGO, the Women's Development Associations Project, to allow families to buy the stove and cylinder sets at market prices. Practical Action worked with the state-owned Nile Petroleum Company (NPC) to facilitate the development of the local LPG market. This involved a new LPG storage tank being commissioned for El Fasher and improvements to the roads leading to the town, which reduced transport costs. Some 12,000 households in El Fasher town have been targeted in each year of the three-year project, and the repayment rate for the loans as of 2015 is 93 per cent.[f] The idea is that any money remaining in the fund will be used to facilitate more energy access initiatives in Sudan or to provide start-up loans for energy enterprises.

[a] See Brendan Bromwich, 'Environmental degradation and conflict in Darfur: implications for peace and recovery', *Humanitarian Exchange Magazine*, Issue 39, Humanitarian Practice Network, July 2008, http://www.odihpn.org/humanitarian-exchange-magazine/issue-39/environmental-degradation-and-conflict-in-darfur-implications-for-peace-and-recovery.
[b] Correspondence with Practical Action Consulting.
[c] Over the life of the programme the exchange rate has varied between $0.4:SDG and $0.16:SDG. The figures in this box use an average exchange rate of ~$0.23:SDG.
[d] Converted on OANDA from SDG for USD at 16 October 2013 rates.
[e] The model assumes LPG at $1.8 per kg. The price given here is around $0.89 per kg.
[f] Practical Action, 'Low Smoke Stoves Project (LPG)', http://practicalaction.org/low-smoke-stoves-project.

The target for lighting is similarly ambitious, with a mix of solar and mini-grid solutions. It is identical to the previous scenario but now at a minimum of Tier 3 access. This consists of at least eight hours energy during daylight hours, with two hours in the evening. It diminishes accident and health risk, and offers people the ability to both light homes and use low-power appliances.[78] Figure 12 shows that the lighting solutions envisaged under the Fundamental Change scenario would require an initial capital investment of $1.31 billion. However, they would save $101 million each year in fuel costs assuming LPG priced at $1.8 per kg.

**Figure 12: Potential savings and capital cost – widespread introduction of LPG and solar and mini-grid lighting solutions**

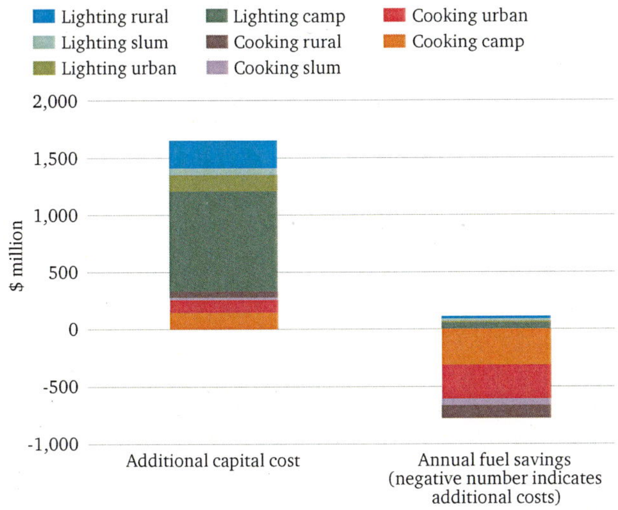


Source: Chatham House Model.

## Reflections on the three scenarios

As the methodology in Box 3 and Appendix A highlights, these scenarios have many limitations. They should be read as a rough approximation of the type of costs and savings involved rather than as a precise calculation. The scenarios take no account of training, maintenance or less than 100 per cent adoption of new technologies. As this report repeats, these factors are all critical to the success of any energy intervention. Furthermore, the models do not account for political or ground-level situations that can support or obstruct any energy intervention.

The three scenarios show that the international community has every incentive to make sure the types of solutions outlined in the **Incremental Change** scenario are realized. The upfront costs of technology are relatively low, and the year-on-year benefits for displaced people massive. The **Alternative Energy** scenario costs more but moves closer to the goal of providing sustainable energy for all. It also marks a significant step towards improving conditions for education and livelihoods. The **Fundamental Change** scenario is more expensive, both in terms of capital costs and annual fuel costs. However, it yields huge potential benefits in terms of safety, health, protection, market generation and many other outcomes.

All of the scenarios incur additional costs and effort unaccounted for in the model. Experience shows that the successful implementation of household energy solutions requires additional supporting measures. These are needed to ensure that equipment is suited to the local context, used correctly and can be serviced and maintained.

Moreover, the Chatham House model in no way suggests that the range of energy interventions outlined in the above scenarios is exhaustive. Other promising avenues exist and should be considered when planning energy interventions. Several camp staff interviewed expressed strong interest in solar cooking and/or communal cooking. Trials in both have had mixed success thus far and data on the experiences of implementing these options need to be gathered so that methods can be adapted for different cultural contexts and food types.[79] The UNHCR estimates that its pilot communal kitchens, used by families in Kutupalong camp in Bangladesh for one meal per day, have cut fuel consumption by a third.[80] One interviewee commented on the potential usefulness of solar cooking options in Tanzania:

> Communal cooking could work very well if we did it for beans. We tried a [solar] 'parabolic' cooker in Nepal and it worked very well. In 10–20 minutes it can cook hot tea. The cookers are shared among four to five families, any of whom can use it in the day. It doesn't cover 100 per cent of use because people will want to cook some things at night and in smaller quantities but for boiling beans, rice, making hot water, it works. And if used for longer than five years [the investment] pays off.[81]

[78] Based on SE4All multi-tier metric assessment for Tier 3 energy use, in scenario 3 all households have Tier 3 energy access for cooking. Azuela, 'Sustainable Energy for All Global Tracking Framework 2015'.
[79] See, for example, Brie Loskota, 'Solar Cooker Project Evaluation, Iridimi Refugee Camp, Chad', October 2007, University of Southern California. Also Bryant Castro Serrato, *Refugee Perceptions Study Za'atari Camp and Host Communities in Jordan*, Oxfam, June 2014, https://www.oxfam.org/sites/www.oxfam.org/files/file_attachments/rr-refugee-perceptions-study-syria-jordan-020614-en.pdf.
[80] Gunning, *The Current State of Sustainable Energy Provision for Displaced Populations*, p. 46.
[81] Interview, UNHCR Country Office, Dar es Salaam, Tanzania, July 2015.

Others expressed interest in experimenting with parabolic stoves for community buildings where water has to be continuously on the boil, not only for hot food and drinks but for sterilization and cleaning. Wind and micro-hydro power are also often cited as having great potential in certain locations.[82]

Perhaps most significantly, the three scenarios do not take into account the potential for change in energy use by humanitarian agencies or government authorities in managing camp facilities, offices and logistics. Tendering for camp-wide services which provide for both these and households could significantly reduce costs by creating economies of scale.

## Further opportunities

Technology costs are falling rapidly as producers find new ways to bring renewable energy to markets, and as rising consumer acceptance encourages markets to expand and develop. Solar panels, batteries and light-emitting diodes have become much cheaper in recent years.[83] Sustainable energy technologies are becoming more competitive, and the best-practice knowledge around implementing them is also improving rapidly.[84] These developments reveal some of the huge opportunities that exist for investing in cleaner and more sustainable energy to produce solutions that have the potential to dramatically improve the situation for all involved. Full reviews of cooking and other technologies can be found in the toolkits that accompany this report.

[82] Private correspondence with Paul McCallion, UNHCR.
[83] The price of solar panels has, for example, fallen by half since 2008. See Jacob Winiecki and Kabir Kumar, *Access to Energy via Digital Finance: Overview of Models and Prospects for Innovation*, CGAP, 2014.
[84] See, for example, GACC, 'The Clean Cooking Catalog', http://catalog.cleancookstoves.org/.

# 4. What is the Basis for New Approaches?

During the research for this report, the desire to manage energy differently has been evident in the perspectives and insights shared by practitioners working in the humanitarian and development space. A shift in thinking is taking place, and this is matched by developments in technology, business models and on-the-ground experience globally. While the pieces are far from being all in place, this chapter surveys the basis and capacity for implementing new approaches that would make energy delivery more sustainable in both camp/non-camp and urban/rural situations.

**A shift in thinking is taking place, and this is matched by developments in technology, business models and on-the-ground experience globally.**

The UNHCR is changing its approach to energy, recognizing that energy access for refugees and others of concern was long neglected. The organization's own reliance on fossil fuel combustion is also under review in an effort to reconcile its energy practices with the UN's commitment to carbon neutrality in its operations by 2020.[85]

Figure 13 illustrates how energy has featured in policy and initiatives in the humanitarian sector over the past 20 years. In May 2014 the UNHCR launched its SAFE strategy for 2014–18, outlined in Box 6.[86] This seeks to enable refugees and other persons of concern to meet their energy needs in a safe and sustainable way.[87] Over the next four years the UNHCR aims to incorporate its energy strategy into related programmes including water, sanitation and hygiene (WASH), shelter, health, nutrition, livelihoods and education. It accepts a need for consistent energy-related data collection and analysis. The aim is to integrate sustainability objectives into its operations to help displaced populations through a number of measures: refugee support and engagement; reduced use of firewood; reforestation; facilitation of access to efficient technology; increased use of renewable energy; and measures to reduce energy demand in households.

[85] UNEP, *Moving Towards a Climate Neutral UN*.
[86] UNHCR, *Global Strategy for Safe Access to Fuel and Energy (SAFE): A UNHCR Strategy 2014–2018*, http://www.unhcr.org/530f11ee6.html.
[87] Strategies have been developed in five countries: Burkina Faso, Ethiopia, Kenya, Rwanda and Chad. Strategies are currently under development in another five countries: Nepal, Uganda, Djibouti, Sudan and South Sudan.

## Engaging private-sector expertise

A growing number of private-sector companies have developed sustainable energy services appropriate for low-income households. This expertise could be harnessed to benefit displaced communities.

## Extending solutions to local populations

Energy solutions for refugee camps could be shared with host countries to boost energy access and security for all.

### Figure 13: Evolution of events and interventions related to energy, and their thematic focus in the humanitarian sector, 1995–2014

| | |
|---|---|
| 1995 | *Boiling Point,* Issue 37. 'Household Energy in Emergency Situations', Stove Checklist for Refugee Situations |
| 1995–ongoing | UNHCR, Environment Unit |
| 1996–2005 | UNHCR, Environmental Guidelines |
| 1997–2011 | Red Cross and Red Crescent, The Sphere Project and Handbook (2004) |
| 1998 | UNHCR, Environmental Guidelines: Domestic Energy in Refugee Situations |
| 2001–2009 | UNHCR, FRAME Toolkit |
| 2002 | UNHCR, Cooking Options in Refugee Situations |
| 2002–2008 | NRC, Camp Management Project: Toolkit |
| 2005 | UNHCR, Forest Management in Refugee and Returnee Situations |
| 2005 | IASC, Clusters creation |
| 2006 | WRC, Fuel and Firewood Initiative: Beyond Firewood |
| 2007 | IASC (UNHCR, WFP, WRC) SAFE |
| 2007 | UNHCR, Handbook for Emergencies |
| 2009 | SAFE Matrix, SAFE Decision Tree Diagrams |
| 2009–ongoing | IUCN, IES, FUEL project |
| 2010 | *Boiling Point,* Issue 59. 'Energy in Conflict and Emergency Relief' |
| 2010 | PA, IRC, EMMA Emergency Market Mapping and Analysis Toolkit |
| 2010 | USAID, Fuel-Efficient Stove Programs in Humanitarian Settings: An Implementer's Toolkit |
| 2011 | UNHCR, Light Years Ahead |
| 2011 | UNIFIL, State of Environment |
| 2012 | WFP, Handbook on SAFE |
| 2012 | UNEP (Post Conflict and Disaster Management Branch), Greening the blue helmets |
| 2012 | Humanitarian Innovation Project |
| 2013 | FAO in Emergencies. Guidance Note |
| 2014 | GACC, Strategy for SAFE |
| 2014 | UNHCR, Global Strategy for SAFE |
| 2014 | UNHCR Energy Lab |
| 2014 | SAFE Humanitarian Working Group website |

Source: Raffaella Bellanca, *Sustainable Energy Provision Among Displaced Populations: Policy and Practice*, Chatham House Research Paper, December 2014, https://www.chathamhouse.org/sites/files/chathamhouse/field/field_document/20141201EnergyDisplacedPopulationsPolicyPracticeBellanca.pdf, p. 7.

New practices are also being trialled on a larger scale as a result of partnerships between humanitarian actors and government and private donors. The UNHCR's partnership with the IKEA Foundation includes the Brighter Lives for Refugees Campaign[88] and substantial support for collecting and improving existing data on refugee energy consumption and needs. The partnership includes a major project in Ethiopia's Dollo Ado camps (which house over 200,000 Somali refugees), aiming at enabling self-reliance and sustainable energy provision. Another partnership, involving the Jordanian government and other financiers, seeks to build a 1-megawatt (MW) solar plant outside Azraq

[88] IKEA Foundation, 'Brighter Lives for Refugees campaign', http://www.ikeafoundation.org/programmes/brighter-lives/.

### Box 6: UNHCR Global SAFE strategy

As part of its commitment to safe access to fuel and energy for its persons of concern, the UNHCR launched its first energy strategy in May 2014. The foundation of the strategy is the vision that 'all crisis-affected populations [should be] able to satisfy their fuel and energy needs for cooking, heating, and lighting in a safe and sustainable manner, without fear or risk to their health, well-being, and personal security'.[a] The strategy is ambitious and comprehensive in scope, taking into account the roles of a wide range of stakeholders and acknowledging the need for appropriate measures for different country and refugee contexts.

The strategy sets the UNHCR and its partners five objectives listed below. These are to:

- integrate energy into emergency readiness and response;
- develop and implement energy strategies at the national level;
- improve access to household fuel and lighting using appropriate technologies and renewable energy;
- increase access to energy for schools, health centres and other institutions; and
- establish and manage woodlots for fuel and provision of environmental protection.

It gives specific attention to the processes and activities needed to achieve these results, including: the need to 'develop methodologies to collect data and use data to make decisions';[b] and to 'capitalize on new and innovative funding opportunities'.[c] There is also a great deal of emphasis on recording experience and applying good practices more widely.

For each strategic objective, the document sets out 'enabling actions'. They emphasize the need to conduct baseline surveys and take into account the present energy situation in camps and their surroundings; and to document experience, measure and evaluate. The first step, and the focus of the 2014–18 strategy, is to support 'priority countries' in developing 'context-specific country programme energy strategies'.[d] This involves training field staff, partners and government counterparts, identifying strategic partnerships for energy, and raising funds.

[a] UNHCR, http://www.safefuelandenergy.org/about/partners.cfm?org=UNHRC, last accessed on 20 October 2015.
[b] UNHCR, *Global Strategy for Safe Access to Fuel and Energy (SAFE)*.
[c] Ibid.
[d] Priority countries identified are Bangladesh, Burkina Faso, Chad, Ethiopia, Jordan, Kenya, Nepal, Rwanda, Sudan and Uganda. Country-specific strategies have been or are being developed for each of these countries, although some of the weaknesses described in this section are replicated in these documents.

camp in Jordan.[89] The plan is to serve both the camp and the three closest villages. These projects are unusual in that they have multi-year funding that allows planning over several years.

The United Nations Relief and Works Agency for Palestine Refugees in the Near East (UNRWA) has no sustainable energy strategy at the time of writing. However, it is collaborating with the Japan International Cooperation Agency to reduce Gaza's overall energy needs through the expansion of the solar industry.[90] This is part of its goal to make Gaza a 'Liveable Place' by 2020. The World Food Programme (WFP) has SAFE projects that require extra-budgetary funding in six countries. The projects include the 'Cooking up Change in Darfur' initiative, which has secured funding from the Dutch Postcode Lottery to introduce fuel-efficient stoves and carry out reforestation in Darfur, Sudan.[91] There are also some innovative partnerships between international and local NGOs that cover displaced as well as local residents, such as the Low Smoke Stoves Project between the UK-based Practical Action, the Sudanese state-owned Nile Petroleum Company and the Darfurian Women's Development Associations Project (see Box 5). This project introduced LPG stoves to displaced people in North Darfur, using carbon financing for the initial capital investment.[92]

Relative to the scale of the task, the UNHCR's SAFE strategy devotes relatively little space to how work and energy measures will be financed over the long term.[93] The strategy's priority is access for populations of concern. It does not look for improvements in the efficiency of institutional energy supply, even though this could result in significant cost savings. The strategy should be read alongside the UNHCR's 2014–18 *Global Strategy for Livelihoods*, which demonstrates the transformation from an institutional approach based on providing for basic needs to one promoting refugee self-reliance. Although it does not mention energy, the *Global Strategy for Livelihoods* promotes the 'right to work and the right to development'. It aims

[89] The idea is to scale this up to 6 MW over time.
[90] UNRWA, *Gaza in 2020: UNRWA Operational Response*, May 2013, http://www.unrwa.org/userfiles/2013070364659.pdf.
[91] WFP, 'Safe Access to Fuel and Energy (SAFE) in Sudan, Information Leaflet 2015', http://documents.wfp.org/stellent/groups/public/documents/newsroom/wfp272054.pdf.
[92] Practical Action, 'Low Smoke Stoves Project (LPG)', http://practicalaction.org/low-smoke-stoves-project, last accessed on 20 October 2015.
[93] UNHCR, *Global Strategy for Safe Access to Fuel and Energy (SAFE)*.

to 'enable people to preserve and protect their productive assets as well as meet their immediate consumption needs'.[94] Much of this strategy is essential if the SAFE energy goals are also to be met and sustained. Relevant aspects include advocacy to allow refugees to work and access land/services; the transition to cash-based assistance; and greater attention to refugee participation in local markets. Further examples of the energy strategies of major humanitarian agencies are set out in Appendix C.

## Lessons learnt on scaling up sustainable energy access

The situations, capacities, needs, aspirations and opportunities for change within displaced and host-country populations differ from context to context. However, it is worth considering some of the general lessons that may be drawn from global efforts to increase energy access over the past two decades.

> The energy access field is benefiting from the proliferation of small to medium-sized technology companies taking risks; and from the spread of mobile phones, mobile money services, smart-grid technology and remote sensing.

Energy product and service supply to base-of-the-pyramid customers in low-income countries is a dynamic sector, particularly where it is decentralized. Promising market-based solutions are developing. Companies and practitioners in the field are learning from and adapting to failure and disruption. The energy access field is benefiting from the proliferation of small to medium-sized technology companies taking risks; and from the spread of mobile phones, mobile money services, smart-grid technology and remote sensing. (The latter can be employed to remotely monitor performance and equipment use with solar home systems, for example.) There is now fairly good knowledge of the constraints involved in making energy services work for poor people.[95]

No magic formula exists, and each case will be affected by local/national conditions, regulations, capacities and culture. However, several success factors appear to apply across the board. These include the need to:

- **Understand the use of energy and its value to displaced people.** How is energy used? What do people want greater energy access for? How much can displaced people pay for energy?
- **Understand what works.** Promising areas include: tariff, service charge or appropriate repayment design; suitable training and capacity-building for end users; and building customer trust through long-term service provider engagement with the market.
- **Encourage local participation and market development.** This can take the form of service management, local supply chain development for equipment and maintenance, and ideally local income-generation activities relating to energy equipment.

Few energy interventions for the poorest in society have been monitored over a significant period,[96] and their proof of concept is thus limited. However, private-sector companies have developed technologies and services addressing funding and operational challenges that could benefit displaced communities. Noteworthy innovations that could be deployed in refugee settings include but are not limited to:

- Rental and lease-to-own (hire purchase) payment terms that could alleviate payment risks for large-scale energy solutions;[97]
- Portable energy infrastructure (i.e. mini-grids, solar farms) whose use can be adjusted by humanitarian or private-sector actors;[98]
- In-camp manufacturing facilities that lower the costs associated with transporting products to remote, insecure locations;[99]
- Remote monitoring infrastructure that allows companies to manage assets and anticipate operations and maintenance issues;[100] and

[94] UNHCR, *Global Strategy for Livelihoods: A UNHCR Strategy for 2014–2018*, http://www.unhcr.org/530f107b6.html.

[95] For a discussion of energy access challenges, initiatives and lessons learned, see Arno Behrens, Glada Lahn, Eike Dreblow et. al., *Escaping the Vicious Cycle of Poverty: Towards Universal Access to Energy in Developing Countries*, Centre for European Policy Studies (CEPS), CEPS Working Document No. 363, March 2012, particularly pp. 10–19, http://aei.pitt.edu/33836/1/WD_363__Behrens_et_al_Energy_in_Developing_Countries.pdf.

[96] However, successful cases such as the IDCOL–Grameen Shakti partnership in Bangladesh have been highlighted in Emma Wilson, Neha Rai and Sarah Best, *Sharing the load: Public and private sector roles in financing pro-poor energy access*, International Institute for Environment and Development (IIED), Discussion Paper, August 2014; also Behrens, Lahn et al., 2012.

[97] GVEP International, private conversation with Bart van Ouytsel, Giertsen.

[98] GVEP International, private conversation with Erwin Spolders, Redavia.

[99] GVEP International, private conversation with David Gerard, Green Bio Energy.

[100] GVEP International, private conversation with George Bowman, Azimuth Power.

- Capacity development for refugees and community members so that they can provide customer education and basic technical services.[101]

Solar energy is spreading without assistance in several places where technology supply chains are established, and where financing is available through remittances or mobile money. For example, a UNHCR study found that some 80 per cent of households in two refugee settlements in southeast Nepal housing around 19,000 people from Bhutan had solar PV panels they had purchased themselves.[102] What works and does not work soon spreads by word of mouth. These evolving findings have relevance for some camp and many non-camp situations of displacement, as the issues are similar for host-country and refugee/displaced populations.

## What does this mean for camp planning?

Human displacement disasters tend to erupt suddenly and unpredictably. The humanitarian response must be mobilized quickly, efficiently and often in challenging geographic situations. In such conditions, little thought is given to the most appropriate energy systems. Instead, tried and tested solutions that are not necessarily the most energy-efficient are put in place. This is the case for both diesel use and the spatial planning of camps.

Thinking beyond immediate refugee needs may open the way for responses such as construction of solar farms or small-scale hydroelectric and water treatment plants, and improvements to water delivery systems. Not only can these interventions reduce energy costs, but they may also bequeath a sustainable benefit to the host country. This can reinforce the public and political acceptability of refugee assistance. Given the availability of training and supply chains, technology applications such as mobile solar or solar-diesel hybrid mini-grids could lower diesel bills over the long term and continue to generate benefits after camps close down. These applications would also be relevant to many off-grid contexts or urban settings in countries where existing infrastructure is overstretched.

### Division of roles and responsibilities

As sustainable energy solutions are a long-term endeavour requiring specialist expertise and potentially legislative and policy support, there is a question over what role UN agencies and humanitarian NGOs are best placed to play. Humanitarian response has tended to be viewed as short-term, and humanitarian agencies are not set up to manage long-term development needs. Out of almost 10,000 staff, the UNHCR has just two dedicated to renewable energy and one full-time economist at its headquarters.[103] It has two technical energy experts in the field in Jordan and Dollo Ado, Ethiopia. These appointments have only been made in the last two years and are not considered permanent positions. At present, humanitarian agencies lack the capacity either to carry out large-scale energy interventions in-house, or to act as a regulator for long-term service contracts.

### Box 7: Allocating risks and responsibilities in managing energy assets using a third-party contractor model

Improving energy access in refugee settings may entail reassigning responsibility for asset performance, optimization and efficiency to dedicated experts – on the grounds that the partners involved would have the qualifications and incentives to operate and maintain energy systems more effectively. This approach could also relieve humanitarian agencies of the responsibility for financing the purchase and installation of new equipment. A number of possible infrastructure management contracts could be tested in an effort to promote effective and efficient energy infrastructure management in displacement settings. Different parties could take the lead in designing, installing/building, financing, owning, operating and maintaining infrastructure assets.

To incentivize operators to optimize asset efficiency and performance, contracts could stipulate payment not on the basis of costs plus profit but on a fee-for-service model. This may reward high service quality or high energy output. Such mechanisms could also allow efficiency gains to be shared between operator and client.

Contract negotiations would force both parties to carefully evaluate the risks in the operation. This process, as well as the accompanying considerations, can be complex and can prompt objections against proceeding. However, given that the status quo itself poses high risks, it would be worth exploring the risks associated with alternatives and how they might be managed. Risks for the management contract model include the longevity of operations, the condition of assets, uncertainty of demand and currency fluctuations. Contractual terms would vary depending on the extent to which different parties were required to finance assets, and the extent to which compensation terms accounted for the associated risks.[a]

[a] Private communication with Ben Good, GVEP International.

[101] GVEP International, private conversation with Henri Nyakarundi, African Renewable Energy Distributor.
[102] UNHCR, 'From Street Lights to Micro-grid, Mission Report', *UNHCR Innovation, Engineers Without Borders*, unpublished, 2015.
[103] Private correspondence with UNHCR HQ staff.

More thought will need to be given to how partnerships for implementing the energy services envisioned within the UNHCR's SAFE strategy will work, and how the UNHCR might retain a high-level governance and regulatory role in camp situations. It needs to ensure that the most vulnerable people have access, yet also delegate responsibilities to other entities for energy delivery to obtain the most appropriate and sustainable results over time. The success of SAFE strategies within the UNHCR, WFP, Food and Agriculture Organization and other agencies will depend on practitioners accepting a growing role for market development, and being open to solutions offered by the private sector. Box 7 looks at how this might work with regard to contracts for camp energy infrastructure management. Essentially, finding the appropriate division of responsibilities between local authorities, humanitarian agencies, development agencies, users and the private sector – and allowing for these relationships to evolve over time – will be key to a project's sustainability.

### Exploring payment for services

Field surveys in Dadaab in Kenya, Goudoubo in Burkina Faso and Dollo Ado in Ethiopia show that many refugee households already pay for energy either in money, time or both.[104] There is a strong business case for allowing private-sector competition in energy service provision, but this must be balanced with core humanitarian principles. The humanitarian sector still relies predominantly on the donation of energy services to displaced people. Such models have often proven counterproductive to development aims because they undermine local markets.[105] In forced displacement settings the case for market-based solutions is not just about handing over responsibility to the private sector. It is also about structuring partnerships that allocate risks to the parties best able to price and manage them, perhaps using public subsidies to make solutions work.

**The humanitarian sector still relies predominantly on the donation of energy services to displaced people. Such models have often proven counterproductive to development aims because they undermine local markets.**

The fact that refugee populations are concentrated in certain locations and supported by international donor agencies offers the opportunity for suppliers to overcome barriers normally associated with the off-grid energy market. These include high manufacturing costs due to low production volumes, high distribution costs arising from the dispersed locations of end users, and limited capital availability to support end-user financing mechanisms.

A combination of the right financing models with support and regulation from ground-level agencies could allow the private sector to more frequently provide lower-cost, safer and more effective energy solutions. In addition, innovative payment mechanisms such as pre-loaded debit cards issued by the UN and cash grants introduced in refugee settings have stimulated local economies and improved refugee self-sufficiency.[106] A 2014 Mercy Corps study on energy in an economically fragile country environment illustrates the difficulties of (and lessons learnt from) setting up and maintaining rural and peri-urban supply chains. For instance, it emphasizes the need for multi-year support for market development programmes and the need for flexibility to adapt these over time, particularly in high-risk situations.[107]

## New thinking in urban settings

Models of cleaner energy supply are evolving for urban settings in low- to middle-income countries hosting large displaced populations. These will provide valuable lessons for policy-makers, camp operators and other stakeholders to share over the next few years. Such models will not necessarily treat displaced people differently from locals and will often bridge the humanitarian/development/environmental sustainability divide. For example, Mercy Corps is working to demonstrate market potential for solar power in Kandahar and Helmand provinces in Afghanistan, where cities are highly dependent on diesel-generated electricity (see Box 8).

In Jordan's Irbid province, the Norwegian Refugee Council (NRC) is introducing renewable energy to help secure shelter and increase the social acceptance of refugees by integrating host-community benefits (see Box 9). Its solar schools project harnesses funding related to climate change mitigation rather than humanitarian aid. It aims to support Jordan's ambition to power state schools with solar electricity.

[104] GVEP International field surveys conducted for the MEI. Also see Haskamp and Haas, *Output 3: Baseline Survey Ethiopia Dollo Ado*.
[105] Bellanca, *Sustainable Energy Provision Among Displaced Populations*.
[106] See, for example, UNHCR, 'Cash grants transform life in Congo camp for Central African Republic refugees', 3 October 2014, http://www.unhcr.org/542eb0ab9.html; or Dale Gavlak, 'Zaatari Syrian refugee camp fertile ground for small businesses', BBC News, 30 July 2014, http://www.bbc.com/news/world-middle-east-28541909.
[107] Mercy Corps, *Illuminating Market Systems Development in Fragile Environments: A Case Study Summary of the Alternative Energy Market in Timor Leste*, 2014, http://www.mercycorps.org.uk/sites/default/files/TimorLesteEnergySummary.pdf.

### Box 8: Pioneering the use of de-risking mechanisms and revolving funds to scale up solar power in Afghanistan

In 2014–15 Mercy Corps has been running a vocational training programme to help integrate unemployed IDPs in the unstable province of Kandahar. This involves courses in trades such as electrical engineering lasting between three and six months. The NGO is setting up a course for solar technicians with electrical engineering as its main subject, finishing with an accreditation in solar energy. The accreditation has been deliberately designed as an 'add-on' to the course. The aim is to avoid introducing excessive numbers of solar engineers into the labour market before there are enough projects available to absorb them, and customers able to pay them.

Meanwhile, Mercy Corps is pursuing schemes to cultivate the market for renewable energy with the support of the Afghan Ministry of Energy and Water. These schemes include demonstration projects and the use of donor funding to help reduce risks for private-sector investors. Projects are structured using lease-sale agreements twinned with technical and business support during the period of each lease.[a]

In an example of a larger project, a 60-kilowatt solar power plant is proposed for the city of Kandahar. The proposal is supported by USAID. 'Wherever there is reliance on diesel generators, there is a case for solar,' says Peter Stevenson, the director of programmes for energy and natural resources in Mercy Corps Afghanistan. 'This is a clearly articulated business model but [it is] amazing how hard it is to be grasped in the NGO community. USAID [has] grasped that.'[b] Kandahar is entirely dependent on electricity generated by diesel and subsidized for the consumer. According to the Afghanistan national electricity company (DABS), the city uses 60,000 litres of diesel a day to power industry and 20,000 litres a day for residential electricity.[c] The US government had been paying around $1 million per month towards that consumption, but reportedly ended this support in September 2015.[d] This means that the Afghan government has significant interest in finding ways to reduce this cost. For development partners to engage in helping to plan for this requires additional donor funding to de-risk private investment.

[a] Mercy Corps is facilitating the construction of a 60-kilowatt solar system serving a Bost University training department in Lashkar Gah, Helmand. This will still be connected to diesel generators as a hybrid system and is expected to yield about 80–90 per cent savings from spring to autumn and 50 per cent savings in winter.
[b] Private communication with Peter Stevenson, Mercy Corps Afghanistan.
[c] Presentation by Eng. Shekib Ahmad Nessar, DA AFGHANISTAN BRESHNA SHERKAT (DABS), 13 June 2015, Kabul, Afghanistan.
[d] USAID Afghanistan, USAID Afghanistan Request for Information (RFI) – Sources Sought Notice: Installation and Supply of 10 MW Solar PV Renewable Energy to Kandahar, issued 9 July 2015.

### Box 9: Using resource-related interventions to work for host communities and refugees in Jordan

The influx of Syrian refugees in urban areas of Jordan has had a pronounced demographic impact.[a] In the North Badya district of Mafraq governorate, the population has risen by some 30 per cent since 2011. In Irbid province, Syrian refugees make up 9 per cent of the population.[b] Most refugees are poor and need support. Some 86 per cent of the refugees in urban areas live below the Jordanian poverty line of $3.20 per day. Public facilities such as schools, water supply and waste collection in these areas are under strain. Channelling aid towards refugee needs in these cases risks exacerbating tensions with local Jordanian residents over resources and humanitarian intervention. This underlines the need to ensure that projects benefit both refugee and host communities.

One example of such an approach involves the International Committee of the Red Cross (ICRC), which has prioritized rehabilitating critical water supply infrastructure (e.g. a water treatment plant, pumping stations and transmission lines) serving a group of urban areas in Mafraq governorate. Services in these areas have been under strain due to the population increase. In the town of Ruwayshid, water is needed for Jordanians and Syrian refugee residents, as well as for Syrian refugees travelling through transit sites in Jordan along the nearby border with Syria. Water demand in Ruwayshid had outstripped supply until the rehabilitation projects were completed in coordination with the Jordanian Ministry of Water and Irrigation and the Yarmouk Water Company. Before the Syrian conflict, some 60 per cent of water in the Mafraq governorate was unaccounted for, much of it lost during conveyance. The ICRC has found that focusing on efficiency in supply and distribution has reduced water losses and increased water supply. As energy is consumed to pump and distribute water, these interventions also cut energy costs.[c]

The Norwegian Refugee Council (NRC)'s solar schools project involves installing solar panels on 20 state schools to reduce costs for the Ministry of Education. Many schools have adopted a system of double shifts – Jordanian children in the morning,

Syrian children in the afternoon. This means energy bills are rising. Solar panels make financial sense in Jordan given the relatively high tariff and government encouragement for solar power. 'A school paying 400–500 Jordanian dinars in electric bills per month would enjoy a payback period of less than four years,' says Annika Hampson, the NRC Shelter project manager in Jordan.[d] The scheme does not use donor money dedicated to humanitarian causes but instead draws on an EU grant for scaling up solar energy in developing countries. The idea is to put the savings into a revolving fund held by the Jordanian Ministry of Education to install more solar systems in state schools. The NRC is also building on earlier success working with local landlords in Irbid province to install solar water heaters in return for secure tenure and rent reduction for refugees.

[a] UNHCR, 'More than four million Syrians have now fled war and persecution', 9 July 2015, http://www.unhcr.org/559d648a9.html.
[b] Salem Ajluni and Mary Kawar, *The Impact of the Syrian Refugee Crisis on the Labour Market in Jordan: A Preliminary Analysis*, International Labour Organization, 2014, http://www.ilo.org/wcmsp5/groups/public/@arabstates/@ro-beirut/documents/publication/wcms_242021.pdf.
[c] Interview with Michael Talhami, ICRC Jordan, March 2015.
[d] Interview with Annika Hampson, NRC Jordan, March 2015.

## Opportunities for aligning energy projects with government policy

As the above examples demonstrate, situations of protracted displacement require approaches and expertise that provide a bridge between traditional humanitarian responses/protection and longer-term development.

Existing government policies and ambitions may provide a framework in which projects and spending to assist displaced people can also contribute to national sustainable development objectives. The new SDGs (see Box 2) reinforce this. Most governments will sign up to these in 2015–16 and incorporate them into their national policies.

Appendix D outlines examples of policies related to energy, forests and refugees in 12 countries hosting large displaced populations. Reforestation is high on government agendas in a number of countries that accept large numbers of refugees. For example, one of the key objectives of Thailand's National Economic and Social Development Plan is to improve environmental quality and continue the 1985 target to expand forest coverage to 40 per cent by 2016. Forest coverage stood at around 32 per cent in 2013 and a new forest protection plan has been launched for the period 2014 to 2023.[108] Uganda's National Forestry Plan aims to restore the country's forest coverage back to 24 per cent of land cover by 2040.[109]

Several countries will offer particular opportunities for collaboration to help meet national energy access and sustainable energy goals. In Nepal, only half of the rural population is connected to grid electricity.[110] Decentralized renewable energy applications have proved important following the earthquakes that hit the country in 2015 and damaged access to power. The country already had a National Rural and Renewable Energy Program 2012–17.[111] Its objectives are to install 25 MW of hydropower and 600,000 solar PV home systems, distribute 475,000 clean cookstoves and provide 130,000 household biogas systems. With experience in off-grid solar applications growing, and the Chinese market close by as a source of relatively cheap panels, use of solar power is expanding in refugee camps in Nepal.[112] Renewable energy use is also increasing in the national energy markets of Kenya and Jordan. These countries' electricity tariffs, combined with policy reforms encouraging renewable energy and energy efficiency, present opportunities for partnerships to build new markets. Kenya has a National Climate Change Action Plan, key aspects of which include clean energy production, introduction of improved cookstoves and reforestation of degraded lands.[113]

In many countries, including Lebanon, Afghanistan and Somaliland, energy generation is dominated by private power providers operating diesel power generators and providing local distribution. This means the cost of electricity is high. In Somaliland, for example, the cost of electricity in 2013 was among the highest in Africa at $1.00–1.40 per kilowatt hour.[114] In these situations, decentralized solar or solar/diesel hybrid systems are often already competitive. An emerging private market may thus offer a basis for investment and projects that aim to scale up these solutions sustainably.

[108] Woranuch Emmanoch, 'Drivers of Forest Change in the Greater Mekong Region: Thailand Country Report', USAID/Royal Forest Department, Thailand, September 2015, http://www.uncclearn.org/sites/default/files/inventory/fao13102015_6.pdf.
[109] Ministry of Water and Environment of Uganda, *The National Forest Plan 2011/12 – 2021/22*, January 2013.
[110] Government of Nepal, 'Scaling-Up Renewable Energy Program, Investment Plan for Nepal, Draft of 11 September 2011', p. 9.
[111] Government of Nepal, 'National Rural and Renewable Energy Programme: Nepal', Programme Document, June 2012, http://www.aepc.gov.np/library/tinymce/uploaded/NRREP%20Programme%20Document-June%202012.pdf, p. 5.
[112] UNHCR, private conversation with Samuel Perkins.
[113] Government of Kenya, *National Climate Change Action Plan: 2013–2017*, http://cdkn.org/wp-content/uploads/2013/03/Kenya-National-Climate-Change-Action-Plan.pdf, 2013.
[114] Ministry of Trade and Investment, *Somaliland: An investment guide to Somaliland Opportunities & Conditions, 2013–2014*, http://somalilandinvest.net/somaliland_investment_guide.pdf, p. 73.

# 5. What are the Challenges to Cleaner, Safer, Sustainable Energy Delivery?

Sustainable energy interventions for displaced people yield massive economic, human and environmental benefits. So why are more of these not already in progress?

Potential solutions need to be introduced with careful attention to the situation on the ground. For the MEI, the choice of technology is not the key question. Instead, the consortium is concerned with what governance and financing structures are needed to make technological and other interventions more accessible and sustainable.

This chapter considers the challenge by breaking it down into three levels of activity. The first is the global/institutional level, where policy, procedures and practice are created.[115] The second is the national level, where these policies and recommendations filter down, are adjusted to the country context, and influenced by host government policy and legislation. The third is the local level, where adverse conditions in the local economy, culture and environment may affect practical implementation.

## Global/institutional challenges

### No institutional home for energy

In 2005 the Inter-Agency Standing Committee (IASC) adopted a cluster-based approach to make the humanitarian response to emergencies more holistic.[116] It states: 'Each cluster lead has accepted to be the agency of "first port of call" and "provider of last resort" within this sector/cluster.'[117] This was intended to bring more accountability to certain key thematic areas.[118] The cluster leads are outlined in Table 4 below.

The wide range of actors involved makes for a confusing division of responsibilities and flows of money. Funding often passes through three or four agencies before reaching its intended beneficiaries.[119]

Although several clusters assist with humanitarian coordination, no cluster is devoted to energy.[120] This means funding for energy projects is rarely given the same level

[115] It should be noted that the subject of 'global refugee policy' itself offers a rich academic discussion around what this constitutes and how it is created. This paper does not have the scope to consider these issues, but for an overview of key debates and theoretical trends relating to global refugee policy, see James Milner, 'Introduction: Understanding Global Refugee Policy', *Journal of Refugee Studies*, Vol. 27, No. 4, 2014.

[116] The cluster system was designed to respond to emergencies in general. It is important to note that not all emergencies are related to forced displacement.

[117] Tim Morris, 'UNHCR, IDPs and clusters', *Forced Migration Review*, Vol. 25, May 2006, http://www.fmreview.org/FMRpdfs/FMR25/FMR2531.pdf.

[118] Abby Stoddard, *International Humanitarian Financing: Review and comparative assessment of instruments*, study for the Good Humanitarian Donorship initiative commissioned by the Office of US Foreign Disaster Assistance, Final report, 22 July 2008, https://www.humanitarianoutcomes.org/sites/default/files/pdf/HumanitarianFinancingReview2008.pdf.

[119] For a more comprehensive review of humanitarian funding, see Stoddard, *International Humanitarian Financing*.

[120] The SAFE Humanitarian Working Group acts as the global coordination mechanism for energy in humanitarian response, but is not a formal part of the UN system. There is also no energy coordination mechanism in-country except in the DRC.

**Short-term policies impede long-term solutions**

**17**

The average length of time as a refugee is 17 years (UNHCR, 2004). In many cases camps' temporary status, maintained for political reasons, inhibits more efficient energy solutions.

**A shortage of data and energy expertise hinders effective interventions**

There is a lack of reliable data on energy costs and use in displacement contexts. To drive reforms, the humanitarian sector needs more dedicated energy experts.

of consideration as issues such as water, sanitation, hygiene and shelter – the emergency responses which automatically draw on dedicated budgets and expertise. As a result, energy responses rarely meet the minimum humanitarian standards for disasters outlined by the Sphere Handbook, *Humanitarian Charter and Minimum Standards in Humanitarian Response*. This specifies that populations affected by disasters should have access to a safe, efficient stove and appropriate means of providing sustainable artificial lighting.[121]

Raffaella Bellanca, in her 2014 review of energy policy and practice in the humanitarian sphere for the MEI, sums up the problem: 'Without an institutionalized space for energy, no one is strongly motivated to advocate for funds and initiatives and no one is truly responsible for the performance of the sector as a whole.'[122]

### Table 4: Inter-Agency Standing Committee cluster leads

| Sector or area of activity | Global cluster lead |
|---|---|
| Agriculture | Food and Agriculture Organization |
| Camp coordination/ management: | |
| IDPs and conflict | UNHCR |
| Disaster situations | International Organization for Migration |
| Early recovery | UN Development Programme |
| Education | UNICEF |
| | Save the Children (UK) |
| Emergency shelter: | |
| IDPs and conflict | UNHCR |
| Disaster situations | International Federation of the Red Cross (Convener) |
| Emergency telecommunications | Office for the Coordination of Humanitarian Affairs/WFP |
| Health | WHO |
| Logistics | WFP |
| Nutrition | UNICEF |
| Protection: | UNHCR |
| IDPs and conflict<br>Civilians (other than IDPs) affected by conflict/disasters | UNHCR/Office of the High Commissioner for Human Rights/ UNICEF |
| Water, sanitation and hygiene | UNICEF |

Source: Based on 'The Clusters' diagram by the UN Office for the Coordination of Humanitarian Affairs, http://www.unocha.org/what-we-do/coordination-tools/cluster-coordination.

## Data and evidence collection

Lack of data collection is evident across the humanitarian sector. For example, the World Food Programme, serving 4.2 million refugees and 8.9 million IDPs, does not appear to collect or publish data on its transport or fuel costs.[123] It lacks a standard procedure for capturing such data.[124] NGO implementing partners collect data in different ways. They may or may not be requested to supply the UNHCR with disaggregated data on fuel costs, supplier fees and generator maintenance costs, for example. There is clear evidence of potential for savings to be made in fuel management and overall energy provision. However, without data collection on the entire system's energy use, with costs disaggregated by service, donors will not be able to make cost–benefit assessments of different energy options. Similarly, private-sector companies cannot propose or implement appropriate solutions.

## Short-term approach to long-term problem

The UNHCR, UNRWA, WFP, many other humanitarian agencies, and the NGOs that work with them are principally organized as emergency relief organizations. They would themselves admit that they are not necessarily the most appropriate or qualified agencies to manage initial urban planning or long-term services such as water and energy infrastructure. In theory their roles should last just months until more durable solutions are found. In many cases, however, relief activities continue for years and sometimes decades without a clear road map from relief to self-sufficiency or consideration of development approaches. Unlike the provision of food and water, which acts as an instant form of relief, energy applications often influence or 'lock in' consumption over years and require regular maintenance. To compound this, host-country governments in many cases fail to acknowledge that camps for displaced people could ever be semi-permanent. Thus there is no political will or incentive to provide long-term services.

The budgeting process within the UNHCR, which is responsible for the majority of camp infrastructures and operational spending, has a number of challenges. The UNHCR is funded on an annual basis by donors. Initial funding allocations to country operations may be reduced when the organization faces funding shortfalls due to, for example, unforeseen emergencies or fluctuations in exchange rates. This short-term funding/budgeting does not lend itself to investments in renewable energy which often take several years to recoup the initial cost and/or generate savings.

[121] Sphere, *Humanitarian Charter and Minimum Standards in Humanitarian Response*, 'Non-food items standard 4: Stoves, fuel and lighting', http://www.spherehandbook.org/en/non-food-items-standard-4-stoves-fuel-and-lighting/, last accessed on 20 October 2015.
[122] Bellanca, *Sustainable Energy Provision Among Displaced Populations*, p. 33.
[123] World Food Programme, 'WFP in Numbers', https://www.wfp.org/wfp-numbers, last accessed on 22 October 2015.
[124] Interview, WFP, August 2015. Data may exist but staff throughout the organization may be unaware of them.

All of the field staff interviewed for this report wanted to reduce diesel costs. Ad hoc plans existed but staff were not generally aware of camp-level or national strategies for diesel cost reduction based on quantitative evidence.

The approach of the International Committee of the Red Cross (ICRC) to infrastructure serving urban areas offers a notable exception as an example of a large international organization trying to bridge emergency and long-term development among displaced populations. The ICRC does not have a mandate regarding refugees but focuses primarily on armed conflict settings (e.g. Syria, Iraq, Yemen) where displacement is also an issue. Over the past 30 years, the ICRC has developed expertise supporting service providers in maintaining and stabilizing essential urban services (water, power and wastewater), thus building relationships with national authorities and technical experts, often in areas subject to recurring conflict. This experience may have relevance for refugee organizations that increasingly need different skillsets and approaches for operating in urban areas. A recent ICRC report finds that the short-term nature of funding is a major obstacle to the kind of long-term infrastructure investment and planning needed in areas of protracted conflict and displacement.[125]

### Procurement

According to the UN Office of Internal Oversight Services (OIOS) audits 2008–15, some challenges faced by the UNHCR's procurement system have included inadequate procurement monitoring by implementing partners, inconsistent procurement policy and application of guidelines, and a lack of training and supervision of procurement staff. These matters are investigated in more detail in one of the accompanying toolkits to this report.[126] However, it should be noted that a new procurement policy was put in place in 2013. As part of this the UNHCR has made significant efforts to optimize processes, decentralize its operations to ensure stronger and more effective management, and improve training processes for supply staff. Today, approximately 65 per cent of the UNHCR's procurement takes place at field level. Toolkits have been distributed to field offices to facilitate local/regional procurement, and several organization-wide strategies have been deployed to improve product selection and the overall tendering approach. Some of the issues highlighted above exemplify the challenges faced by humanitarian actors more generally.

Despite reform efforts, several concerns remain. Procurement processes in humanitarian organizations most often relate to equipment rather than final energy services. Of 18 United Nations Global Marketplace energy tenders between 1 February and 1 September 2015, 83.3 per cent have been requests involving specific technologies.[127] Thus the tendering requests may be said to preselect a technology rather than encourage technology-neutral solutions by asking for responses to the actual energy need. The process of procurement decentralization has brought its own problems, since selected product suppliers have not always been of sufficient quality.[128] Finally, private companies interviewed as part of the research for this project also raised questions about the standards that their products must comply with. One interviewee said that these standards were often viewed by the sector as 'wholly inadequate or mismatched with the actual need'.[129] NGOs working in the field do not necessarily apply the same technology standards. Too much inconsistency restricts the ability of companies to customize products for the sector, invest in appropriate research and development, and bring down costs. To attract the most practical solutions, NGOs and agencies in the sector should seek a balance between uniformity of standards and openness to innovation/adaptability to local context.

### Problems of scale

Pilot projects using more energy-efficient equipment or renewable energy have often failed because they did not consider local social conditions, preferences or market structures. In too many cases the quality of the evaluation work has been weak. The focus has tended to be on how many items of equipment have been distributed rather than whether the equipment is successful in establishing sustainable energy systems. In many cases successful pilot projects have not been scaled up because funds have run out. Others have stalled due to a lack of financial support or a failure to properly engage the local community – which is often essential to ensure effective management and maintenance. For example, the MEI survey in Goudoubo, Burkina Faso showed that only 1 per cent of households surveyed used a solar cooker and only as a secondary cookstove.[130] Box 10 below outlines another example of the difficulties of such initiatives.Table 5 gives a selection of pilot projects mentioned in our interviews with camp staff.

[125] ICRC, *Urban services during protracted armed conflict: a call for a better approach to assisting affected people*, 6 October 2015, https://www.icrc.org/en/document/urban-services-protracted-conflict-report.
[126] Lindsay Van Landeghem, *Private-Sector Engagement: The Key to Efficient, Effective Energy Access for Refugees*, GVEP International Toolkit for the MEI, 2015.
[127] United Nations Global Marketplace, https://www.ungm.org/Public/Notice.
[128] Lindsay Van Landeghem, *Private-Sector Engagement: The Key to Efficient, Effective Energy Access for Refugees*, GVEP International Toolkit for the MEI, 2015.
[129] Private correspondence with established private-sector producer of off-grid solar solutions.
[130] Practical Action, field survey in Burkina Faso, 2015.

### Table 5: Selected examples of pilot sustainable energy projects in refugee camps

| Pilot location and date | Type | What was achieved? | Funder and implementer | Experience |
|---|---|---|---|---|
| Kutupalong, Bangladesh 2008 | Lighting | 160 solar street lights installed. | UNHCR, Rahim Afroz installed the street lights. | Only 124 of the street lights are still functional, many have deteriorated and technology is old. |
| Sag-Nioniogo, Burkina Faso, 2013 | Cooking | Solar cookers ('blazing tubes') and solar lanterns distributed. Construction of mud stoves introduced. Reached 691 households. | UNHCR | Households gave a mixed reaction to the 'blazing tubes'. These are said to be still working well in some contexts but are confined to daylight hours so this limits use. The mud stove materials were too cumbersome and proved not to be durable during the rainy season. |
| Abala, Niger, late 2013 | Cooking | Gas project covering 2,500 households (10,000 refugees). Three stone stoves are connected to 6-litre and 12-litre bottles. The stoves are made locally, and host population is included in distribution. | UNHCR, French embassy to Niger provided funding for 14 months, EU expected to fund an additional 10,000 households in 2015. Sonihy, a local Niger company, is acting as a supplier. | 40 per cent of targeted population still uses wood for some cooking, but the project is generally seen as successful and has reduced impact on environment. Project is being expanded and is actively seeking further funding. |
| Emkulu, Eritrea, 2014 | Cooking | Introduced improved energy-saving stove; 100 households were targeted, 10 women were trained to build them, two days were needed to build them. The cost of one stove is $120 so cannot be covered by households. | UNHCR funded by Swiss embassy in Khartoum. | The UNHCR deemed the pilot successful but there were difficulties finding further funding to continue the project. The initial donor chose to support non-energy projects rather than continue with the same pilot. |
| Kawergosk, Iraq, 2015 | Shelter | Improved shelter including solar panels to provide energy to power a light/ charge a phone; 12 units were supplied. | UNHCR, IKEA Foundation – Better Shelter | Lack of funds currently prevents pilot expansion. |

Sources: Interviews with UNHCR field staff; GVEP International and Practical Action site surveys; SAFE Project pages, SAFE, 'Where We Work', http://www.safefuelandenergy.org/where-we-work/index.cfm.

One of the problems with pilot projects is that by definition they are small-scale. However, the evidence suggests that a key barrier to off-grid energy market formation is the need to achieve scale to drive down delivery costs. By designing small-scale pilots and not planning for scale-up, the humanitarian community is missing an opportunity. There is a clear need for longer-term commitments to large-scale projects. If these take shape, then improved delivery systems can be exploited to provide cost-effective services to local communities as well as displaced people.

## National challenges

### Translating policy into practice

All projects need the support of host governments. This can be impeded by a number of general problems in the relationship between governments and humanitarian agencies. On the one hand, during a refugee crisis host governments can be overwhelmed by multiple international agencies with varying levels of aptitude for service delivery and varying understanding of local governance systems. Lack of coordination of relations with governments is a common problem. Many states will be reluctant to surrender functions considered 'governmental', such as power provision and camp planning, to humanitarian agencies. UN agencies often face tensions in dealing with government bureaucracy. Negotiations to secure even limited rights to work or access land for agriculture or gain permission

> **Box 10: Santo 17, Crois des Bouquets, Port-au-Prince, Haiti**
>
> During the humanitarian response to the 2012 earthquake in Haiti, several international organizations decided to collaborate to produce a showcase for renewable energy supply in displaced contexts. They installed a biogas system in Santo 17, a relocation camp in Haiti hosting 358 families. It consisted of five biodigesters and a mini-grid powered by five units, each comprising eight PV panels and eight batteries. The project fell into disarray due to poor planning of accountability structures, inadequate assessment of the socio-cultural context and technical failures. The biodigesters never functioned properly, and the street lights went out after a few months.
>
> Adapted from Raffaella Bellanca, *Sustainable Energy Provision Among Displaced Populations: Policy and Practice*, Chatham House Research Paper, December 2014, https://www.chathamhouse.org/sites/files/chathamhouse/field/field_document/20141201EnergyDisplacedPopulationsPolicyPracticeBellanca.pdf.

to upgrade camp shelters can take years. There is a fine line to tread between defending the interests of refugees and maintaining constructive relations with host governments.[131]

### Duration of stay

There is usually great political sensitivity concerning refugees' length of stay in the host country. This may limit the scope for optimal infrastructure planning that could save energy over the long term, and for any interventions that could make a settlement look more permanent. The UNHCR typically aims for wider community access to energy in both host and refugee contexts. Its policy is to pursue alternatives to camp settings wherever possible while ensuring that refugees are protected and assisted effectively.[132] Settlements for displaced people are rarely planned for the long term, but in reality they tend to exist for a minimum of five years, and there are at least 21 cases of settlements continuing for more than 20 years.[133]

### Legal arrangements

Access to energy opens up many opportunities for forcibly displaced people, yet prohibitive legal arrangements restricting freedom of work and movement often inhibit the ability of refugees to make the most of these opportunities (see Box 11). Refugees' levels of self-reliance and needs vary widely. Some camps become informal towns or slums with their own markets and services. The Sahrawi refugee camps in the Western Sahara are examples. They have a growing tourist industry and relative autonomy in terms of service distribution.[134] Many others, such as the Dadaab camps, remain highly dependent on aid and the

### Box 11: Major differences in economic rights for displaced people

People forced to leave their countries are accorded widely varying rights to freedom of movement, employment, and access to land and public services. For example, no national asylum framework exists in Thailand, which is not a signatory to the 1951 Refugee Convention or 1967 Protocol. This limits the rights of forcibly displaced people entering the country, who are classified as 'persons of concern'. Their treatment is often unregulated. Burmese refugees are considered illegal if found outside the designated camps for Burmese asylum-seekers. This restricts their ability to seek and keep jobs.

Uganda is at the other end of the spectrum. Refugees there 'have the right to work, to move freely within the country, and to live in the local community rather than settlements'.[a] Camp residents in Nakivale, established in 1958, have been allotted access to agricultural land which they can farm and whose produce they can sell – although certain crops are restricted.

In most non-camp situations, at least one person in each refugee household works informally if work permits are not issued. For example, it is well known that refugees in Lebanon and Jordan without legal permission to work accept wages much lower than the norm. These two countries have the highest ratios of refugees to nationals in the world.[b] Permits may vary from camp to camp. In Jordan, for example, the Zaatari camp allows economic activity within its confines, and a thriving market place has emerged. This has led to local perceptions of Zaatari residents as relatively affluent by the standards of some local Jordanians who endure poor living conditions. In response to the concerns this created, the Azraq camp was set up without grid electricity. Moreover, economic activity by refugees in the camp was initially forbidden, with shops and businesses reserved for Jordanian owners and workers. However, in 2015 negotiation brought some relaxation of this rule, with new shops to be operated on a 50/50 basis by Jordanian nationals and refugees.

Many countries have few or no laws governing refugee rights. The rules that apply to refugees are ad hoc and heavily influenced by the political and economic environment. Where displacement is protracted, the UNHCR and other humanitarian actors often engage in discussions with the host government to confer greater rights on displaced populations. A change in their status may sometimes come about after many years of careful planning and political debate.[c] This was the case for 163,000 refugees of Burundian origin in Tanzania in 2014. They were naturalized and are now allowed to work outside their settlement and move to other parts of the country.

[a] Naohiko Omata and Josiah Kaplan, 'Refugee livelihoods in Kampala, Nakivale and Kyangwali refugee settlements: Patterns of engagement with the private sector', Working Paper Series No. 95, October 2013, Refugee Studies Centre, University of Oxford.
[b] *UNHCR Global Trends 2014: World at War*, p. 3.
[c] Interview with country office staff, Dar es Salaam, Tanzania, July 2015.

[131] Private correspondence with local non-governmental agency working in South Asia.
[132] UNHCR Policy on Alternatives to Camps, http://www.unhcr.org/5422b8f09.pdf, 2014.
[133] Palestinian refugee camps (1948): Rafah, Khan Younis, Nuseirat, Jabalia, Beach, Amman New Camp, Ein El Hilweh; Tanzanian camps set up in 1972: Katumba, Mtabila; Mae La, Thailand (1984); Dadaab (Kambios, Hagadera, Dagahaley and Ifo), Kenya (1992); Coopers Camp, West Bengal (1947); Beldangi, Nepal (1990); Buduburam, Ghana (1990); Nyarugusu, Tanzania (1996); Kutupalong, Bangladesh (1994); Mayukwayukwa, Western Province, Zambia (1966); Navikale, Uganda (1958).
[134] For further details and nuance on this oft-quoted example, see Elena Fiddian-Qasmiyeh, *The Ideal Refugees: Gender, Islam, and the Sahrawi Politics of Survival*, Syracuse University Press, 2014.

distribution of basic supplies. Hundreds of thousands of displaced people outside official camps, such as the Rohingya in Bangladesh, lack any official status. They remain highly vulnerable, with poor shelter and high rates of malnutrition.

**Enterprises such as agroforestry become possible where refugees are permitted to work, for example in Nakivale in Uganda or in the South Sudanese camps. This reduces stress on the local environment from energy use, and increases refugees' potential ability to pay for energy services.**

Enterprises such as agroforestry become possible where refugees are permitted to work, for example in Nakivale in Uganda or in the South Sudanese camps. This reduces stress on the local environment from energy use, and increases refugees' potential ability to pay for energy services. Other rights need to be secured with local communities. There may be opportunities for energy-related funding to play a role in this respect. For example, refugees and local communities in Pariang county in South Sudan received fuel-efficient stoves and benefited from training in stove construction. This helped build trust between the two groups and generated an agreement allowing refugees to access the river for fishing.[135]

## Local challenges

Each refugee settlement is different. To be sustained, energy solutions designed and introduced by external agents must be accepted and used by refugee populations.

### Social acceptability

Energy pilot projects have often failed where social and cultural contexts have not been taken into account. For example, the use of biogas for cooking was rejected by groups in Somalia who are not comfortable using energy produced from human waste.[136] Similarly, communal cooking solutions need careful design to ensure they are consistent with local cultures. Some large-scale cooking of staple foods has been successful, such as with the *injera* (flatbread) cooked communally in some refugee camps in Ethiopia.[137] Another example is rice cooked in solar 'parabolic' cookers in Nepal. In some places, outdoor communal cooking has led to problems when neighbouring families have joined meals uninvited to share scarce food.[138] In spite of the benefits in terms of air quality, briquettes are often rejected because people dislike the smell.[139] Another objection to briquettes is that traditional dishes do not taste the same as when cooked with wood or charcoal.[140]

### Instability and insecurity

Places where refugees have settled are often politically unstable and conflict-affected. During the research for this project, the relief efforts of several humanitarian providers were seriously obstructed by fighting and the siege (in progress since 2012) in Yarmouk, the Palestinian settlement outside Damascus. Humanitarian agencies were also forced to abandon the al-Obaidi refugee camp in central Iraq due to the threat from Islamic State of Iraq and Syria (ISIS). Similar problems seriously impede humanitarian efforts in dozens of locations around the world. In such places the difficulties of protecting workers and facilities would obviously be major concerns for any party considering investing in public infrastructure, including energy facilities. Even in established refugee settlements, armed escorts can be required. Where this is the case, this adds enormously to operational costs and can and make it hard for camps, agencies, implementing partners or businesses to run normally.

### Geographical location and climate

The physical geography and climate of the displacement setting make a difference. Refugees and displaced people are often settled on land of poor agricultural quality. This makes subsistence or communal farming difficult or impossible, and often inhibits livelihood activities. This is as true for urban refugee populations as for those in camps. Box 4 highlights an example in South Sudan, where physical geography makes the response to refugee crises more difficult. Refugees are often grouped in remote areas, making it harder to

[135] Interview with camp staff, South Sudan, 2015.
[136] Private correspondence with Paul McCallion, UNHCR, 2014.
[137] Notably Mai-Aini. See UNHCR, 'Ethiopia Refugee Update, January-February 2013', http://www.unhcr.org/513da76f9.pdf.
[138] GVEP International Chad assessment, 2015.
[139] Boris Atanassov, 'Sustainable Biomass Briquette Pilot', GreenLight Projects, 2013, http://english.rvo.nl/sites/default/files/2014/02/Additional%20deliverable%20from%20DBM%2002045%20-%20Briquette%20Pilot.pdf.
[140] Practical Action, field survey in Burkina Faso, 2015.

transport essential items to them. This provides further justification for decentralized energy solutions, which present significant advantages and reduce dependence on diesel-powered transportation.

The design of renewable energy solutions, decisions on energy-related adaptation and, in both cases, the associated choice of materials must take climate and weather patterns into account. For example, many camps in Africa use plastic roofing, which disintegrates in the hot sun. This has to be replaced every two years or so – another expense relying on fossil fuels while not providing adequate insulation against heat. Knowledge of local conditions is also essential for projecting the energy savings from any given solution and selecting the right technology for the situation. In areas where water is scarce, biogas from camp waste may not be a viable option. For example, 27 small biogas plants were trialled in 2013 in Nayapara, a Rohingya camp in Bangladesh where waste is a big problem. This did not work as expected due to the scarcity of water needed to flush the waste to the plant.[141] With regard to cooking, procurers, implementers and trainers involved in the distribution of fuel-efficient cookstoves often lack prior knowledge of the moisture content of locally available firewood. This significantly affects the stoves' efficiency in use.

### Competition among NGOs and other stakeholders

Local and national NGOs and civil society representatives can find themselves squeezed out by the arrival of international organizations. Tensions may arise on the ground between local implementing partners and national and international partners. Dinçer et al. have observed many of these dynamics in the Turkish response to the Syrian refugee crisis.[142]

### Economy

Local economies are often intertwined with camp economies. Local villagers sometimes use camp facilities such as schools and clinics, just as refugees will often buy goods in local markets. Trade in charcoal, firewood and in some cases diesel, kerosene or LPG can form a vibrant part of the camp/local economy. For example, a UNHCR official in Tongogara, Zimbabwe told us: 'We don't encourage production of charcoal but the refugees have some barter with the host community, and charcoal is one of the preferred commodities … there is some charcoal production just outside camp.'[143] Charcoal production and distribution can run as a commercial activity in many camps. It may happen illegally, but the groups involved will often have a strong interest in maintaining their market. Other monopolies or cartels often arise around contracts for fuel for camps. The interest groups concerned may oppose energy sustainability plans that envisage reducing or eliminating the use of certain fuels.

Regardless of the presence of such opposition, any plan that significantly affects supply chains will need to consider the impacts on local livelihoods. Many sites hosting refugees are in rural border areas with harsh conditions of poverty. In these cases, inadequate supply chains and difficulties with 'last mile' delivery of services to customers can be a constraint for energy provision, as can limited overall investment in economic development.

[141] Cultural and economic issues also played a role in this. In the other refugee camp in Bangladesh, UNHCR officials reported that the biogas system is working well.
[142] Osman Bahadır Dinçer et al., *Turkey and Syrian Refugees: The Limits of Hospitality*, Brookings Institution, 2013.
[143] Chatham House interview with UNHCR officer in Zimbabwe.

# 6. Conclusions and Recommendations

The sharp rise in the numbers of people displaced by conflict is putting unprecedented pressure on the humanitarian system and on countries providing refuge. This report has examined just one area of humanitarian response: energy. If managed better, energy services have the potential to ease some of these pressures and benefit both displaced people and host countries. This should not in any way detract from the urgent need for political, economic and environmental solutions to prevent displacement or to hold to account those responsible for repression and persecution. Given the scale of today's refugee crises and the long-term work needed to solve their causes, there is a strong case for increasing financial support and improving the effectiveness of humanitarian spending.

The authors find that energy services to displaced people fall far short of what is needed. The current lack of adequate heat, light and power affects millions of the world's most vulnerable people, who consume vast amounts of firewood and charcoal for cooking and kerosene for lighting. This situation is contrary to the principles of humanitarian assistance. The authors also find that energy use in humanitarian operations themselves, such as refugee camp facilities, transportation and offices, is largely unaccounted for. This means that resourcing implications are not fully captured in the data, and that the opportunity to introduce cleaner, more sustainable and cost-effective solutions is missed.

Developing countries host 86 per cent of forcibly displaced people worldwide. Many nationals from these countries already suffer the effects of severe energy poverty, as well as environmental degradation, pollution, water scarcity and deforestation.

As a top priority, we call for the global agenda for energy access and sustainable energy to include forcibly displaced populations. The particular conditions and problems associated with sustainable energy poverty to these groups must be acknowledged.

Second, we argue that the time to act is now. The humanitarian sector is willing to reform while the private sector, social enterprise and NGOs have built a wealth of experience in providing poor people with access to sustainable energy. These advantages must be harnessed. However, the management and funding of energy will have to change to ensure success in displacement situations. At present, the short-term funding cycles of humanitarian agencies limit the scope for investment in longer-term, energy-efficient, low-carbon projects. Agencies also lack the expertise and resources to plan for and manage energy effectively on a large scale. To a great extent, the answer is both to increase in-house expertise and outsource more services to the private sector, using careful regulation to ensure accountability for performance over time.

In view of these findings, we list a number of recommendations below for host governments, humanitarian agencies and donors, and specific suggestions to help achieve them.

### 1. Incorporate sustainable energy access for displaced people into international, national and agency agendas.

**Displaced people should not be excluded from the energy access agenda.** The World Humanitarian Summit is approaching, and the International Decade of Sustainable Energy for All is entering its third year. Actors from international, national and local organizations should thus ensure that the energy needs of forcibly displaced people are appropriately represented in these global programmes. There would be greater progress if countries hosting large displaced populations explicitly recognized their needs within national energy access targets. This report recommends that 10 countries recognize displaced populations within national SE4All energy access targets by the end of 2016.

**Humanitarian agencies should incorporate energy considerations into core programming for each stage of a humanitarian response.** Adequate planning for energy supply and services should begin prior to an emergency, be a feature of the immediate response, and continuously be integrated into the response as the situation develops. Energy provision does not need to be treated as a priority at every stage of the humanitarian response, but checks should be in place to ensure best practice is followed. The adoption of global SAFE strategies by leading humanitarian organizations is a progressive development in this respect, particularly in terms of management of household energy use in refugee camps. Sustained implementation will require the appointment of permanent head office experts on renewable and non-renewable energy. Contingency planning and coordination, which has begun under the SAFE Humanitarian Working Group, should continue to be integrated into all relevant operational sectors among implementing partners. A further mechanism should be established to prioritize institutional set-up of camps, office operations and logistics. All relevant humanitarian agencies should have access to a regularly updated best-practice guide on energy service provision in different contexts.

**More attention should be given to the provision of energy services in urban settings.** This is likely to require quite a different approach. The expanding majority of displaced people will live in or on the edges of cities. This means energy services in these settings deserve significant attention, especially as a means of protecting and assisting with integration.

### 2. Build the data.

**The humanitarian sector needs to collect and report disaggregated data on fuel use, energy practices and costs.** Detailed data on energy use, equipment efficiencies and costs, local entrepreneurship models, the availability of capital for refugees and potential payment mechanisms should be collected in a standardized manner, particularly for refugee camps. It is also critical that energy and efficiency data for institutional administration and operations – including logistics and supplier fees – are incorporated into accounting. The data can be used to demonstrate the costs and benefits of energy interventions and to provide a basis for competitive tendering. The priority will be to specify and define the necessary data requirements, and to develop practical processes for effectively capturing and analysing the data. For example, the UN agencies are well placed to set the data reporting requirements for their implementing partners in camp settings. SAFE or environmental representatives can be required to collect these data and submit them to headquarters. More work is needed to establish processes for capturing better information on energy provision in non-camp situations.

### 3. Coordinate national ambitions and humanitarian aims for mutual benefit.

**Energy access and other resource sustainability challenges should be considered areas for cooperation between host countries, international donors and humanitarian agencies.** Where forcibly displaced populations are unlikely to return home within a short time frame, energy and water projects offer an opportunity for humanitarian aid to contribute to longer-term country development. At the same time, these can reduce the national costs of hosting refugees. If a government subsidizes the fuel costs charged to humanitarian agencies, both will also have a joint interest in scaling up sustainable energy applications to cut fuel demand over time. Accepting that most refugees stay in the host country for at least five years clarifies the business case for capital investments likely to generate fuel cost savings after two or three years.

**Large-scale energy interventions must support local and national sustainable energy strategies.** Countries hosting refugees have ambitions to increase the sustainability of their energy systems and often to increase energy access for their own populations. Energy interventions will have the greatest chance of being accepted and supported if they aim to support these national goals. Wherever possible, strategies should be developed that benefit host as well as displaced communities. Roles and responsibilities for implementation should be clearly laid out over time. For example, any large renewable energy plant established to serve a refugee camp should also serve the local community. In particular, projects relating to public infrastructure shared with local residents, and to housing owned by locals, should be examined as a potential means of offering sustainable benefits and developing local markets. Schools, health facilities, and water and sewerage systems are examples.

**Relaxing national restrictions on freedom of movement and work, such as those banning refugees from earning a living, will facilitate sustainable energy services in many countries.** Restrictions on working prevent long-term refugees from returning benefits to the host state. Similarly, forbidding refugees from engaging in official transactions like paying for electricity connections leaves them vulnerable to higher diesel costs or illegal (and often dangerous) electricity connections. It forces transactions into the informal economy, which in turn denies the host government potential tax revenues that could go towards funding humanitarian support for forcibly displaced populations.

### 4. Embed energy projects and accountability at the local level.

**The design of energy interventions must take into account the needs and capabilities of displaced and local communities.** Projects should make the most of opportunities to actively engage with the communities they serve through training in installation, maintenance and supply chain management. This means mapping economic linkages between refugee and host communities, local laws and regulations, the capacities of local energy service providers, the characteristics of local energy markets, and the geographic and climatic factors affecting technology selection.

**Improve communication channels between ground staff and humanitarian agency headquarters.** Field operations, refugees and their representatives, and external researchers should be allowed to supply feedback on energy delivery to headquarters and propose change. This could significantly improve the effectiveness and sustainability of energy delivery strategies.

**Ensure accountability for the performance of energy interventions lies with providers and implementers, with a framework over several years for evaluating their sustainability.** For example, businesses may consider renewable technologies a one-off sales opportunity, but where there is low capacity to maintain and manage facilities the model needs to be one of service provision over time. The question should be how best to allocate and manage risks and responsibilities for energy asset ownership, financing, operations and maintenance to the benefit of all parties involved.

**Encourage responsibility among staff for their personal energy usage, and for the energy usage of their programmes and projects.** There are many ways to cut costs and emissions. Staff can be incentivized to support these efforts while not compromising on the quality of services and comfort levels provided. Several measures could be introduced. These could include encouraging the use of mass transport to and from camps (already available in some places), smart metering in camp offices and redistribution of energy savings to other essential items or services.

### 5. Explore new delivery models.

**Support for displaced people needs to move away from a model based on handouts.** The majority of finance flows in during the initial phase of an emergency and then soon tails off, leaving large fuel and electricity bills that are increasingly difficult to pay. The focus of the current model for serving the needs of refugees and displaced people is to provide free items and rations. This is essential in the early stages of a crisis and for the most vulnerable households. However, initial emergency relief should transition to more sustainable energy provision centred around refugee self-reliance. This will mean paying much greater attention to refugee energy access and payment mechanisms, and to the evolution of sustainable market dynamics. Services based on cash or vouchers could play a significant role in this transition.

**Recognize that cleaner energy delivery extending beyond initial emergency funding requires expertise beyond the humanitarian field.** Mechanisms are required to encourage and maintain in-camp energy interventions. These should be allocated to actors with the capacity and embedded incentives to ensure that energy solutions operate effectively over their designated time frames. Examples of such mechanisms include consumer education, facilitation of distribution, provision or management of financial mechanisms, after-sales service, and training in use and maintenance. Recognizing that this expertise will not always be available within the humanitarian field would open the way for more innovative tendering and public–private partnerships. This would aid the introduction of business models better able to support long-term sustainable energy services. Given the fragile and temporary nature of many refugee settlements, such partnerships would benefit from including local stakeholders.

**Take steps towards overhauling procurement practices and standards for equipment and services related to energy.** Procurement processes in the humanitarian sector should become less concerned with specific equipment and more concerned with the nature of services needed in each context, in order to promote the most appropriate technology and performance accountability over time. The transparent development of clear, uniform standards should draw on input from relevant private-sector providers and groups that share the relevant level of technical understanding. Procurement should not simply specify the cheapest solution on the global market, but should consider each proposed solution's carbon footprint and potential impact on local market development. In terms of equipment and technologies, the sector should adopt clear, uniform standards for product performance and quality wherever possible while leaving room for innovation and adaptation to local contexts. Tools are already available, such as the GACC's Clean Cooking Catalogue and the IFC's Lighting Global Quality Standards.[144] Host communities should play a stronger role in the procurement process, which should offer them a greater range of choices. These adjustments will not only be relevant to energy, but also to other programming areas such as water, sanitation, hygiene and shelter.

**Develop contract frameworks for camp infrastructure management in consultation with relevant development and private-sector experts.** Management of large-scale energy assets – which in refugee settings are mostly purchased, operated and owned by humanitarian actors – could be competitively outsourced. This would enable appropriately qualified and incentivized expert teams to take responsibility for asset performance, optimization and efficiency. It would also relieve humanitarian agencies of the responsibility for financing the purchase and installation of new equipment. The idea would be to shift the model from paying for *assets* to paying for *services*. For example, incentives could be geared towards increasing the efficiency and utilization of renewable energy sources.

**Dedicate time and funds to experimenting with different approaches to assessing needs, planning, and using more flexible implementation and distribution methods.** Donors should direct funds to build on pilot projects that have already proved successful. Doing this at a larger scale, such as for a whole camp, would allow energy providers to achieve cost and operational efficiencies. Expertise can be significantly improved as experience is gained and communicated throughout the humanitarian system.

[144] Global Alliance for Clean Cookstoves, 'The Clean Cooking Catalog', http://catalog.cleancookstoves.org/; and the International Finance Corporation's Lighting Global Quality Standards, https://www.lightingglobal.org/qa/standards/.

### 6. Explore innovative funding models.

**Where possible, encourage the use of local markets to sustain and cultivate energy solutions.** In many cases a basis for cleaner energy solutions may already exist. Innovation will be needed to connect such options to the displacement setting, which may be preferable to introducing completely new technology. Solutions based on local markets can also generate income by drawing on the entrepreneurial talent of both displaced populations and host communities. During sourcing, it is advisable that administrators consider commercial proposals that will establish effective and sustainable distribution channels. For example, these could include franchising opportunities for competitively selected pay-as-you-go solar companies, local solar kiosks/service centres or cookstove distribution enterprises.

**Stress test mechanisms to allocate risks between humanitarian agencies, governments and private-sector parties.** Such mechanisms should allow investments to be made outside the normal public-sector funding cycle and framework. For example, donors should allow their funding to be deployed in the form of soft capital to de-risk private-sector investment and kick-start local markets and supply chains. It should be possible to devise or source insurance that protects asset purchasers against risks associated with refugee settings.

**Consider linking large-scale camp energy infrastructure contracts to opportunities to expand services to nearby households.** A significant infrastructure management contract will motivate a contractor to relocate to the vicinity of a given camp. This can create the opportunity to service cost-effective 'bolt-on' obligations to expand energy access to nearby households – potentially among both the displaced population and the host community. The contract will need to be carefully designed to ensure the correct performance and efficiency incentives. It will need to allocate risks in a way that both meets client needs and attracts compliant bids during procurement.

## Next steps

Drawing on the above recommendations, the MEI consortium members plan to work together with relevant governments, policy-makers, NGOs and donors to:

**Establish a high-level panel** to build support and funding for scaling up successful solutions globally. This would be composed of influential organizations and experts working in energy access, development, poverty alleviation, sustainable energy, climate resilience and humanitarian relief. The panel would bridge gaps between different sectors of expertise and sources of potential financing. It would be capable of reaching and conversing with government decision-makers in host and donor countries.

**Create and manage a fund** that makes loans to agencies proposing to invest in projects. It could be structured as a revolving fund with the participation of donor countries, international development finance institutions and possibly philanthropic foundations. The fund would be separate from other contributions to the UNHCR or other agencies. Existing operational budgets would not be affected. It could also be designed to incorporate a supplementary technical assistance facility. This would help those proposing new projects to identify and develop investment proposals. This would require thorough research and a pipeline of diverse projects. The application of standards for project finance would also help improve data by demanding baseline assessments and performance audits.

**Design a detailed model for camp planning to assist with energy and related decision-making.** This would ask the right questions to enable planners of refugee camp energy interventions to understand the impact of different choices. Projections made at the outset of a crisis based on real data in established camps could inform decision-making on all energy infrastructure and equipment. This could include appropriate types of shelters, camp layout, lighting, cooking, heating and cooling equipment. This could become an open-source facility helping to promote innovation for the purposes of broader problem-solving in the energy and displacement context.

**Establish a private-sector humanitarian network forum.** This would enable private-sector firms interested in providing energy services to displaced populations to learn of opportunities to engage effectively with humanitarian organizations. Relevant parties could then jointly consider introducing and/or scaling up energy solutions in refugee settings. One of the forum's key activities would be to develop or agree on and harmonize progressive technology standards for equipment across the humanitarian system. This could be coordinated with the Safe Access to Fuel and Energy (SAFE) Humanitarian Working Group, in partnership with independent technical experts and industry associations. It could help open up the market place, increase innovation and cut costs.

**Pilot site-specific integrated energy plans in several large displacement contexts.** The aim would be to create a blueprint for meeting targets on energy access, efficiency and carbon emissions at each designated site. Implementation would rely on completing the necessary groundwork, which would consist of several activities. First, a detailed energy audit of infrastructure assets would assess their performance and efficiency, and the opportunities for optimization/fossil fuel reduction. Second, there would be a review of the needs and preferences of refugee and local communities. Third, local stakeholders would need to be carefully consulted. The process should take into account vested interests, previous work and especially the opportunities to create benefits for host communities either through improved energy access or enterprise/market development. Once the assessments had been carried out, pilot projects would concentrate on providing large-scale energy solutions with greater outsourcing of implementation to the private sector than has been attempted before. This would mean using a camp infrastructure management contract. Pilot projects would have to ensure that monitoring and evaluation took place over several years to draw out lessons learnt.

**Research, design and pilot non-wood concessions at scale.** Even efficient cookstoves do not solve the problems associated with firewood collection or harmful emissions. They also do not necessarily pass WHO indoor air pollution safety guidelines. Yet there is little capability to test non-wood solutions at scale. Efforts to promote non-wood fuel interventions have noted that a one-stove solution does not fit all. Cookstoves must, therefore, be tested to prove their applicability and feasibility in large-scale initiatives. Promoting large numbers of non-wood fuel interventions would require building on lessons learnt from some of the pilots already completed – e.g ethanol/LPG trials in Ethiopia and Niger. The main objective should be to create a tender for a large non-wood concession with all the necessary standards. It would need both to bring down overall costs and significantly reduce or eliminate reliance on firewood in a camp or area over a period of years.

# Appendix A: Methodology behind the Chatham House Model

## Overall purpose

In its first phase the MEI has sought to understand and evaluate energy use by displaced populations around the world both within and outside camps. Using a purpose-built model, the MEI can estimate the scale and cost of present energy use and $CO_2$ emissions among displaced households, as well as the cost of potential interventions. This model does not evaluate the energy use by camp administration and operations, although this important topic is considered separately in the main body of this report.

The model incorporates the costs of different fuels (firewood, charcoal, kerosene etc.) for cooking and lighting. It also covers consumption levels and the capital costs required to provide basic energy equipment (stoves, lanterns etc.). Heating is not considered explicitly in the model. However, the use of fuel for heating is considered to be largely synonymous with fuel use for cooking.[145] Prices, capital costs and consumption figures used by the model are based on interview data and independent research. Details are available on request from the authors. All aspects of the model can be adjusted in line with new data as these are fed into the project.

**Table A1: Price assumptions**

| | Prices for cooking fuels in $/kg |
|---|---|
| Charcoal | 0.18 |
| Firewood | 0.07 |
| Processed solid fuel | 0.18 |
| Biogas | 0.00 |
| LPG | 1.80 |
| Kerosene | 1.28 |
| Biomass briquettes | 0.23 |

| | Electricity unit costs for lighting fuels |
|---|---|
| Torches | $4/month |
| Kerosene | $1.5/litre |
| Electricity | $0.03/kWh (includes capital costs) |
| Mini-grid 1 | $0.01/kWh (running costs only) |
| Mini-grid 2 | $0.03/kWh (running costs only) |
| Diesel generator | $0.05/kWh |
| Solar | $0.00 |

The methodology behind the MEI's model was presented and discussed during an international expert roundtable in London on 18 June 2015. The detailed technical assumptions behind the cooking and lighting assumptions were analysed and verified by independent experts from the Global Alliance for Clean Cookstoves (GACC) and SolarAid.

## Interviews

The MEI began by conducting 24 semi-structured telephone interviews with specialist UNHCR staff across the globe. These interviews lasted approximately one hour each. A list of key questions that had been compiled by the consortium was circulated to interviewees in advance. Interviewees were encouraged to respond to these set questions, but were also free to tell interviewers about unique or individual circumstances arising within the settlements where they were working. The interviews provided a rich array of qualitative and quantitative data, but also revealed large gaps in the overall information available.

## Camp populations

Based on the data from these interviews and independent research on displacement contexts, the MEI developed a set of baseline energy use patterns for refugee households. Average figures for fuel consumption, energy costs and $CO_2$ emissions were calculated for each baseline type.

The typology includes five baseline types for cooking in camps, and four baseline types for lighting in camps.

## Cooking and lighting typologies

The tables below set out the baseline types and targets for cooking and lighting. Baseline types describe a pattern of energy use among forcibly displaced people in particular kinds of settlements. Targets describe an improved pattern of energy use in terms of tier level of access and fuel use. In effect, they describe how the baseline settings would change if specific improvements were introduced in the relevant households/communities. The relationship between types and targets is described in more detail below.

As mentioned, the average fuel consumption figures set out below were derived from interview data and independent research. Of necessity these are highly generalized averages. In reality there is considerable

[145] Many displaced people, for example, rely exclusively on the warmth of the cooking fire for heating. Costs of heating fuel are also often beyond the capacity of many displaced populations. In contexts where expenditure on fuel for heating is particularly significant, such as in parts of western Asia and Europe, the use of fuel explicitly for heating is worthy of further investigation.

### Table A2: Baseline and target cooking types[146]

| | Type | Minimum tier | Description | Charcoal | Firewood | Processed solid fuel | Biogas | LPG | Kerosene/ other |
|---|---|---|---|---|---|---|---|---|---|
| | | | | Baseline energy use types and targets (kilogrammes per household per month) | | | | | |
| Baseline | 1 | 0 | Firewood-dependent | 2.44 | 119.16 | 0.00 | 0.00 | 0.07 | 0.34 |
| Baseline | 2 | 0 | Firewood-charcoal mix | 14.84 | 70.46 | 0.00 | 0.00 | 0.15 | 0.00 |
| Baseline | 3 | 0 | Liquid | 2.44 | 43.89 | 0.00 | 0.00 | 1.77 | 11.11 |
| Baseline | 4 | 0 | LPG | 13.00 | 36.00 | 0.00 | 0.00 | 4.41 | 0.00 |
| Baseline | 5 | 0 | Alternative biomass | 0.00 | 65.49 | 55.18 | 5.21 | 0.00 | 0.00 |
| Target | 1 | 3 | Firewood-dependent | 1.83 | 41.70 | 0.00 | 0.00 | 0.07 | 0.26 |
| Target | 2 | 3 | Firewood-charcoal mix | 11.13 | 24.67 | 0.00 | 0.00 | 0.15 | 0.00 |
| Target | 3 | 3 | Liquid | 1.83 | 15.36 | 0.00 | 0.00 | 1.77 | 8.33 |
| Target | 4 | 3 | LPG I | 9.75 | 12.60 | 0.00 | 0.00 | 4.41 | 0.00 |
| Target | 5 | 3 | Alternative biomass | 0.00 | 22.92 | 26.21 | 2.47 | 0.00 | 0.00 |
| Target | 6 | 3 | Biomass briquettes | 0.00 | 16.30 | 37.07 | 0.00 | 0.00 | 0.41 |
| Target | 7 | 3 | LPG II | 0.00 | 0.00 | 0.00 | 0.00 | 13.40 | 0.00 |

### Table A3: Baseline and target lighting types[147]

| | Types | Minimum tier | Description | No access (%) | Torches (%) | Kerosene (%) | Grid electricity (%) | Mini-grid 1 (%) | Mini-grid 2 (%) | Household diesel generator (%) | Solar (%) |
|---|---|---|---|---|---|---|---|---|---|---|---|
| | | | | Lighting baseline types and targets (share of population using fuel) | | | | | | | |
| Baseline | 1 | 0 | Torch-dependent | 18 | 61 | 4 | 0 | 0 | 0 | 10 | 7 |
| Baseline | 2 | 0 | Kerosene-dependent | 5 | 20 | 60 | 5 | 0 | 0 | 0 | 10 |
| Baseline | 3 | 0 | Electricity-dependent | 0 | 5 | 5 | 90 | 0 | 0 | 0 | 0 |
| Baseline | 4 | 0 | Solar-dependent | 5 | 10 | 20 | 0 | 0 | 0 | 0 | 65 |
| Targets | 1 | 1 | Solar/diesel | 0 | 0 | 0 | 0 | 0 | 0 | 50 | 50 |
| Targets | 2 | 1 | Grid | 0 | 0 | 0 | 100 | 0 | 0 | 0 | 0 |
| Targets | 3 | 1 | Solar/mini-grid | 0 | 0 | 0 | 0 | 50 | 0 | 0 | 50 |
| Targets | 4 | 3 | Grid | 0 | 0 | 0 | 100 | 0 | 0 | 0 | 0 |
| Targets | 5 | 3 | Solar/mini-grid | 0 | 0 | 0 | 0 | 50 | 0 | 0 | 50 |

variation in fuel consumption between different displacement contexts. Firewood consumption, for example, is contingent on the type of wood, type of stove and climate. In cases where firewood is used in inefficient fires, estimates of baseline fuelwood use vary from 0.7 kg per person per day up to 3 kg per person per day.[148] Our Tier 0 estimate of 119.16 kg of firewood consumption per household per month in firewood-dependent households is at the lower end of that spectrum.

## Population

Data on the location and size of uprooted populations were taken from the statistical annexes to *UNHCR Global Trends 2014: World at War*.[149] This dataset encompasses a population of 49,053,874 displaced people sorted by type of location (individual accommodation, undefined/unknown, collective centre, reception/transit camp, self-settled camp and planned/managed camp). A distinction was made between

[146] For more detailed information on these consumption patterns and the data they were based on, please contact the authors.
[147] For more detailed information on these consumption patterns and the data they were based on, please contact the authors.
[148] Gunning, *The Current State of Sustainable Energy Provision for Displaced Populations*.
[149] *UNHCR Global Trends 2014: World at War*, Tab 15 of Annexes, http://www.unhcr.org/pages/49c3646c4d6.html.

camp (collective centre, reception/transit camp, self-settled camp, planned/managed camp) and non-camp populations (individual accommodation, undefined/unknown). Based on this distinction, the camp population comes to 8,696,922 and the non-camp population to 40,356,952 of the 49,053,874 in the original dataset.

### Camp populations

Each camp with a population over 20,000 was individually assigned a baseline energy use type. This was determined both by interviews with UNHCR staff and by independent research on those camps.

Camps with a population of under 20,000 were grouped by regional location (sub-Saharan Africa, Middle East [western Asia] and North Africa, Asia, Europe, Central Asia and South America). Proportions of the displaced populations in those regions were then allocated to baseline types according to the proportion of camp populations over 20,000 in their region allocated to each type. Thus if Cooking Type 2 was assigned to 20 per cent of the inhabitants in Central Asian camps with over 20,000 people, it was also assigned to 20 per cent of inhabitants of camps with under 20,000 people.

For both cooking and lighting, Type 1 is by far the most prevalent among camp populations globally.

### Non-camp populations

Non-camp populations provide additional methodological dilemmas since they are often widely dispersed throughout the host country. This implies a wide variety of contexts and energy use patterns. With the exception of Jordan, the study did not analyse the energy use of non-camp populations within a particular country.

Non-camp populations were sorted by country (according to the original UNHCR dataset) and then by location within that country (urban, rural, slum). The distinction between urban and rural locations was made on the basis of Global Tracking Framework (GTF) data on urban ratios in each country. The evidence available suggests that forcibly displaced people are disproportionately urban, so a weighting was applied to the urban populations to ensure that this cohort was sufficiently represented.[150]

The urban refugee populations were then split again to create a differential between 'urban' and 'urban slum'. The data on which we based this division were drawn from national UN-Habitat data.[151]

National GTF data were then used to highlight levels of access to solid and non-solid cooking fuels and to grid connections for both urban and rural populations. Slum populations were assigned an average of the urban and rural rates for grid access, and the same rates as urban populations for access to non-solid fuels. For households with access to non-solid cooking fuels, LPG was assumed to be the fuel burnt, with IEA country data used to calculate household LPG consumption.[152] For households without access to non-solid fuels, a baseline energy use type was assigned in line with the camp model (Type 2, firewood/charcoal – for urban populations; Type 1, firewood-dependent – for slum and rural populations). For those connected to the grid, IEA country-level consumption data were used to calculate costs.[153] Displaced households without grid connection were categorized as having Type 1 baseline energy use for lighting (dependent on torches or less).

## Calculations

For total fuel spend figures, the number of displaced households in a camp (or in a country, for non-camp populations) was multiplied by the annual fuel costs per household for its type.[154] Thus the number of Type 1 households was multiplied by the average baseline fuel cost for a Type 1 household. In non-camp contexts the number of displaced households assumed as on-grid or with access to non-solid fuels was multiplied by each cohort's respective spending in terms of national on-grid kWh per year per household,[155] based on IEA data.[156] These individual annual cost figures were then summed to give the $2.1 billion total annual fuel spend figure cited earlier in this report.

The total $CO_2$ emissions figures were calculated in a similar way, multiplying the number of households by the annual emissions (in $tCO_2$) per household for its respective type. The figures for each household type and for grid-connected households were totalled, to give the 13 million $tCO_2$/year figure cited elsewhere in this report.

150 In *UNHCR Global Trends 2014: World at War*, a proportion of the 87.3 per cent of refugees in private individual accommodation are described as living in urban settings. We used this proportion as a weighting to scale up the displaced urban populations in each country, so that the correct proportion would be represented globally. While this proportion only refers to refugees, no better proportions were found for the urban–rural split for IDPs, for example.
151 UN-Habitat, 'Proportion of Urban Population Living in Slums', United Nations Statistics Division, 2007, http://urbandata.unhabitat.org/explore-data/?indicators=slum_proportion_living_urban,population,urban_population_cities,hiv_prevalence_15_to_49_year.
152 IEA, *Energy Statistics of OECD Countries 2014*; IEA, *Energy Statistics of Non-OECD Countries 2014*.
153 Ibid.
154 Due to a lack of widespread data on the number of households in camps and non-camp settings, we assumed five people per household across all displaced contexts.
155 The proportion of households on grid and with access to non-solid fuels was calculated on the basis of GTF data as outlined above.
156 IEA, *Energy Statistics of OECD Countries 2014*; IEA, *Energy Statistics of Non-OECD Countries 2014*.

## Scenarios

The model allows the MEI to estimate the impact of different interventions on both the annual energy consumption and $CO_2$ emissions of displaced households. It also allows the MEI to chart the projected capital cost of such interventions. Currently the model outlines three potential target scenarios for cooking and lighting (see Table A6). These targets vary in their ambition. Although the cooking and lighting targets are independent of each other, they are described together for ease of reference.

Tables A2 and A3, above, set out the patterns of energy use for each type and target. Tables A4 and A5 set out the shifts from a baseline type of energy consumption to a target type of energy consumption under each of our three scenarios.

### Table A4: Shifts from type to target under three cooking scenarios

| Baseline | | Scenario | | |
|---|---|---|---|---|
| Type | Description | Incremental Change | Alternative Energy | Fundamental Change |
| | | Target | Target | Target |
| 1 | Firewood-dependent | 1 | 6 | 7 |
| 2 | Firewood/charcoal mix | 2 | 6 | 7 |
| 3 | Liquid fuel | 3 | 3 | 7 |
| 4 | LPG | 4 | 4 | 7 |
| 5 | Alternative biomass | 5 | 5 | 7 |

### Table A5: Shifts from type to target under three lighting scenarios

| Baseline | | Scenario | | |
|---|---|---|---|---|
| Type | Description | Incremental Change | Alternative Energy | Fundamental Change |
| | | Target | Target | Target |
| 1 | Torch-dependent | 1 | 3 | 5 |
| 2 | Kerosene-dependent | 1 | 3 | 5 |
| 3 | Electricity-dependent | 2 | 2 | 4 |
| 4 | Solar-dependent | 3 | 3 | 5 |

In the first target cooking scenario – that is, **Incremental Change** – camps and non-camp populations retain their baseline patterns of energy use but become more efficient, moving up tiers. For example, households in a Cooking Type 1 (firewood-dependent) camp will continue to use the same proportion of firewood as other fuel types (98 per cent firewood, 2 per cent charcoal). However, they will consume less of that fuel in kilogrammes per month as they adopt more efficient stove types (moving from Tier 0 to Tier 3 stoves, for example).[157] In broad terms, the scenario therefore describes the widespread introduction of improved cookstoves. The **Incremental Change** target lighting scenario describes a situation where all uprooted populations have better access to energy for lighting. Camps and populations that were previously dependent on torches (Lighting Type 1) and kerosene (Lighting Type 2) all move to adopt a 50/50 split between solar and diesel use for lighting at minimum Tier 1 access (Lighting Target 1). Camps or populations whose baseline type is dependent on electricity (Lighting Type 3) already have a large degree of grid connectivity. These move to 100 per cent grid connectivity at a Tier 1 access level (Lighting Target 2). Camps previously dependent on solar (Lighting Type 4) will move to adopting a 50/50 split between mini-grid and solar at Tier 1 access (Lighting Target 3).

In the **Alternative Energy** scenario, biomass briquettes (Cooking Target 6) are introduced into baseline firewood-dependent (Cooking Type 1) and firewood/charcoal mix (Cooking Type 2) camps and populations. In this scenario just over two-thirds (69 per cent) of fuel use shifts to briquettes at a Tier 3 access level. Biomass briquettes are an increasingly common fuel source in developing countries, often made from agricultural waste, recycled materials or other materials such as sawdust. The other three cooking types retain their original consumption patterns but at higher tiers, as set out in the Incremental Change scenario. For lighting the Alternative Energy scenario describes Lighting Types 1, 2 and 4 shifting to a 50/50 split between mini-grid and solar at a minimum Tier 1 access (Lighting Target 3). Lighting Type 3 populations and camps remain at 100 per cent grid connectivity, as in the Incremental Change scenario (Lighting Target 2).

The third target scenario – **Fundamental Change** – for cooking describes the widespread introduction of LPG (Cooking Target 7) across all camp types and populations. For lighting the scenario envisages Lighting Types 1, 2 and 4 once again shifting to a 50/50 split between mini-grid and solar, this time at a minimum Tier 3 access level (Lighting

[157] Each household will consume 41.70 kg of firewood per month rather than 119.16 kg per month. This can be seen in Table A2 by comparing Type 1 fuel use and Target 1 fuel use.

Target 5). Lighting Type 3 populations and camps again shift to 100 per cent grid connectivity, but also at Tier 3 rather than Tier 1 access (Lighting Target 4).

Table A6 below summarizes the general shift in energy use patterns in each scenario for cooking and lighting.

### Table A6: Table outlining three potential targets for cooking and lighting

| | Incremental Change scenario | Alternative Energy scenario | Fundamental Change scenario |
|---|---|---|---|
| Cooking scenarios | Baseline energy use patterns retained but with more efficient equipment | Introduction of biomass briquettes in baseline Types 1 and 2 | Widespread introduction of LPG |
| Minimum tier level | 3 | 3 | 3 |
| | **Incremental Change scenario** | **Alternative Energy scenario** | **Fundamental Change scenario** |
| Lighting scenarios | Access to basic lighting for all, primarily through solar and diesel use | The majority of refugees connected to mini-grid, with solar also featuring prominently | The majority of displaced people connected to mini-grid, with solar also featuring prominently |
| Minimum tier level | 1 | 1 | 3 |

Other solutions not incorporated within our target scenarios may involve the use of a combination of locally produced biogas, more communal or commercial cooking methods, and solar cookers. However, the lack of studies proving the effectiveness of these technologies at scale means that we have excluded them from any scenario in this report.

## Limitations

The model is the first attempt to estimate the scale of global energy use in situations of forcible displacement. This model has a number of limitations. It is a simplified model of a highly complex system. Our typology does not, for example, adequately cover the vast array of different patterns of energy use in existence. It relies heavily on basic proxies for energy use within particular countries. Patterns of energy use in displaced households will vary both within and between camps, but our model only takes the latter into account. The model thus gives only an indication of the kinds of numbers we are dealing with. It should not be taken as a comprehensive picture of energy use among forcibly displaced people. The data used in the model both for assigning types and for cost and consumption assumptions are imperfect. The data based on interviews were often patchy, aggregated and unverified. The interventions described in the target scenarios are necessarily meant to be indicative of the costs and benefits of certain approaches, and are not an endorsement of a particular approach. The model does not account for a less than 100 per cent adoption of a given intervention, and only accounts for basic technology costs. It excludes the crucial costs of distribution, maintenance, market creation etc. The model also excludes the significant political, technical and geographical constraints which may make certain interventions unlikely or even impossible.

# Appendix B: Table of Definitions

| Term | Definition |
|---|---|
| Persons of concern to the UNHCR | A term used to describe all people whose protection and assistance are of interest to the UNHCR. This includes refugees, asylum-seekers, IDPs, returnees and stateless persons under the UNHCR mandate (54.9 million by end of 2014).[158] |
| Refugee | A person who 'owing to a well-founded fear of being persecuted for reasons of race, religion, nationality, membership of a particular social group or political opinion, is outside the country of his nationality, and is unable to, or owing to such fear, is unwilling to avail himself of the protection of that country.' Regional instruments have since extended the definition to other categories of people.[159] |
| Internally displaced persons (IDPs) | Internally displaced persons are persons or groups of persons who have been forced or obliged to flee or to leave their homes or places of habitual residence, in particular as a result of or in order to avoid the effects of armed conflict, situations of generalized violence, violations of human rights or natural or human-made disasters, and who have not crossed an internationally recognized state border.[160] |
| Urban | De facto population living in areas classified as urban according to the criteria used by each area or country.[161] (GTF based on IEA data, which in turn rely on UNDESA Population Division, World Urbanization Prospects.) |
| Rural | De facto population living in areas classified as rural according to the criteria used by each area or country.[162] (GTF based on IEA data, which in turn rely on UNDESA Population Division World Urbanization Prospects.) |
| Slum | The proportion of the urban population lacking at least one of the following five housing conditions: access to improved water, access to improved sanitation facilities, sufficient living area, structural quality/durability of dwellings, security of tenure.[163] (UN-Habitat) |

[158] *UNHCR Global Trends 2014: World at War*, p. 8.
[159] Individuals recognized under the 1951 Convention Relating to the Status of Refugees, its 1967 Protocol and the 1969 Organization of African Unity Convention Governing the Specific Aspects of Refugee Problems in Africa.
[160] UNHCR, 'Guiding Principles on Internal Displacement', http://www.unhcr.org/43ce1cff2.html.
[161] UN Department of Economic and Social Affairs (UNDESA), Population Division, *World Urbanization Prospects, the 2014 revision*, 'Glossary of Demographic Terms', http://esa.un.org/unpd/wup/General/GlossaryDemographicTerms.aspx.
[162] Ibid.
[163] UN-Habitat, 'Proportion of Urban Population Living in Slums', UN Indicators Methodology Sheets, http://www.un.org/esa/sustdev/natlinfo/indicators/methodology_sheets/poverty/urban_slums.pdf.

# Appendix C: Energy Policies of Major Humanitarian Agencies

| Organization | Sustainable development strategy/policy/activity | Energy efficiency/sustainable energy initiatives |
|---|---|---|
| International Committee of the Red Cross (ICRC) | In September 2011, ICRC adopted a sustainable-development policy with the objective of incorporating environmental protection into its operations and decision-making processes.[164]<br>Framework for Sustainable Development[165]<br>Framework for environmental management in assistance programmes[166] | ICRC delegations[167]<br>Biogas projects: Prisons in the Philippines (2011),[168] Nepal, Rwanda[169]<br>Solar projects: Philippines (2015),[170] Jordan,[171] Gaza,[172] Uganda (2012),[173] South Sudan[174] |
| International Organization for Migration (IOM) | IASC Task Force SAFE Standing Invitee[175] | Partner on SAFE Project to distribute stoves in Haiti 2010[176]<br>Partner on SAFE Project solar light distribution and impact survey in IDP camps Haiti (2012–13)[177] |
| International Federation of Red Cross and Red Crescent Societies (IFRC) | SAFE Steering Committee Member[178]<br>IASC Task Force SAFE Standing Invitee<br>IFRC Approach to sustainable development 2011 Position Paper[179] | |
| Mercy Corps | Mercy Corps is committed to increasing access to clean energy through its programmes.[180]<br>Mercy Corps Energy for All Programme[181] | Solar projects: Timor-Leste,[182] Uganda (solar market assessment)[183]<br>Clean cookstoves: Uganda,[184] Myanmar[185] |
| Médecins Sans Frontières (MSF) | Research Partnership with Veolia Foundation aimed at reducing MSF's energy footprint and examining opportunities for renewable energy provision for field facilities.[186] | |

164 ICRC, 'World Environment Day: ICRC committed to sustainable development', 6 April 2012, https://www.icrc.org/eng/resources/documents/feature/2012/sustainable-development-feature-2012-06-04.htm.
165 ICRC, *Sustainable Development at the ICRC Annual Report*, 2013, https://www.icrc.org/en/document/sustainable-development-icrc-annual-report-2013.
166 ICRC, 'The three dimensions of sustainable development at the ICRC', https://www.icrc.org/eng/assets/files/2012/sustainable-development-f-and-f.pdf.
167 ICRC offices and premises in Amman, Abidjan, Beijing, Bogota, Geneva, Harare, Juba, Mexico, Monrovia, Nairobi, New Delhi and Paris have undergone sustainable development programmes. Private correspondence with ICRC staff.
168 ICRC, 'Waste no more: introducing renewable energy in Philippine jails', 25 March 2011, https://www.icrc.org/eng/resources/documents/feature/2011/philippines-feature-2011-03-25.htm.
169 ICRC, 'Ensuring Appropriateness of Biogas Sanitation Systems – Analysis from Rwanda, Nepal and the Philippines', 1 July 2012, https://www.icrc.org/eng/resources/documents/article/other/biogas-article-2012-07-01.htm.
170 ICRC, 'Philippines: Eco-friendly project in Davao City Jail', 19 May 2015, https://www.icrc.org/en/document/philippines-eco-friendly-project-davao-city-jail.
171 Private correspondence with ICRC staff.
172 Private correspondence with ICRC staff.
173 ICRC, 'Uganda: When a solar panel helps save lives', 2 April 2015, https://www.icrc.org/en/document/uganda-when-solar-panel-helps-save-lives.
174 ICRC, 'World Environment Day: ICRC committed to sustainable development', 6 April 2012, https://www.icrc.org/eng/resources/documents/feature/2012/sustainable-development-feature-2012-06-04.htm.
175 Inter-Agency Standing Committee Task Force on Safe Access to Firewood and alternative Energy in Humanitarian Settings, 'Decision Tree Diagrams on Factors Affecting Choice of Fuel Strategy in Humanitarian Settings', 2009, https://www.humanitarianresponse.info/system/files/documents/files/Decision%20tree%20diagrams%20on%20factors%20affecting%20choice%20of%20fuel%20strategy.pdf.
176 SAFE, 'OCHA/ERRF and IOM Stove Distribution', http://www.safefuelandenergy.org/where-we-work/project.cfm?p=37.
177 SAFE, 'WakaWaka Light distribution in Haiti', http://www.safefuelandenergy.org/where-we-work/project.cfm?p=111.
178 SAFE, 'International Federation of Red Cross and Red Crescent Societies', http://www.safefuelandenergy.org/about/partners.cfm?org=IFRC.
179 IFRC, 'The Red Cross Red Crescent approach to sustainable development', Position paper, 2011, https://www.ifrc.org/PageFiles/90969/The%20Red%20Cross%20Red%20Crescent%20approach%20to%20sustainable%20development,%20April%202011%20(English)%20-email.pdf.
180 Mercy Corps, 'Environment', https://www.mercycorps.org.uk/environment.
181 Energy for All, 'Mercy Corps Saves and Improves Lives in World's Toughest Places', http://www.energyforall.info/organisation/mercy-corps/.
182 Mercy Corps, 'Life of a girl: Solar power lights the way for Lourdes', 10 August 2015, https://www.mercycorps.org.uk/photoessays/timor-leste/life-girl-solar-power-lights-way-lourdes.
183 Mercy Corps, 'Mercy Corps Market Analysis for Household Solar Products in the Acholi Sub-Region', Uganda, 2015, https://www.mercycorps.org/research-resources/mercy-corps-market-analysis-household-solar-products-acholi-sub-region-uganda.
184 Mercy Corps, 'Making clean energy affordable far from the city', 30 October 2013, https://www.mercycorps.org.uk/articles/uganda/making-clean-energy-affordable-far-city.
185 Mercy Corps, 'Innovative conservation efforts honored', 31 July 2012, https://www.mercycorps.org.uk/articles/myanmar/innovative-conservation-efforts-honored.
186 Veolia Group, 'Médecins Sans Frontières (Doctors Without Borders) and the Veolia Foundation sign research and innovation partnership', press release, 16 March 2015, http://www.veolia.com/en/veolia-group/media/press-releases/medecins-sans-frontieres-doctors-without-borders-and-veolia-foundation-sign-research-and-innovation-partnership.

| Organization | Sustainable development strategy/policy/activity | Energy efficiency/sustainable energy initiatives |
|---|---|---|
| World Food Programme (WFP) | 2009 SAFE Initiative – In 2009 WFP committed to reach 6 million displaced persons through a combination of activities addressing the issues WFP beneficiaries face when cooking.[187]<br>Leading member of SAFE Steering Committee<br>IASC Task Force SAFE Co-Chair | SAFE programmes: Ethiopia, Haiti, Kenya, Sri Lanka, Sudan, Uganda, Burundi |
| Food and Agriculture Organization of the United Nations (FAO) | The FAO is committed to responding to the cooking needs of assisted populations in emergency and recovery contexts.[188]<br>IASC Task Force SAFE Member<br>Leading member of SAFE Steering Committee | Darfur Timber and Energy Project[189]<br>Solar Cooling Project Angola[190]<br>Strengthening linkages between refugee and host communities in Kakuma to improve incomes, food security and ultimately nutrition.[191] SAFE activities in Ethiopia, Kenya, Somalia, South Sudan and Myanmar |
| UNICEF | IASC Task Force SAFE Member<br>Greening UNICEF Policy[192] | UNICEF Burundi Innovations Lab (2012)[193] |
| Women's Refugee Commission (WRC) | Co-Chair of the SAFE Initiative[194]<br>WRC has been working since 2005 to put cooking fuel on the humanitarian agenda.<br>Spearheaded creation of IASC Task Force and the SAFE Initiative. | |
| United Nations High Commissioner for Refugees (UNHCR) | Global Safe Energy Strategy 2014–2018[195]<br>Leading member of SAFE Steering Committee[196]<br>UNHCR Environmental Guidelines[197]<br>UNHCR Energy Lab[198]<br>UNHCR, The Environment and Climate Change[199] | Light Years Ahead rolled out in Chad, Ethiopia, Kenya and Uganda[200]<br>FRAME Toolkit[201]<br>UNHCR SAFE strategy rolled out in Burkina Faso, Chad, Ethiopia, Kenya, Nepal, Rwanda, Sudan and Uganda 2014–15[202] |

[187] WFP, 'Safe Access to Fuel and Energy (SAFE) Initiative', https://www.wfp.org/safe.
[188] FAO, 'Safe Access to Firewood and alternative Energy sources (SAFE)', http://www.fao.org/energy/78520/en/.
[189] FAO, 'UNEP and Partners in Sudan Joint Programme on Environment and Natural Resources Management: Darfur Timber and Energy Project', http://www.fao.org/emergencies/fao-in-action/projects/detail/en/c/195682/.
[190] FAO, 'Solar PV systems, solar fridges and cold store rooms', http://www.fao.org/energy/82416/en/.
[191] FAO, 'FAO and MasterCard announce new partnership', 2 September 2015, http://www.fao.org/news/story/en/item/326780/icode/.
[192] UNICEF, 'Greening UNICEF', http://www.unicef.org/environment/index_65215.html.
[193] UNICEF, 'The Right to Energy', 1 March 2015, http://www.unicefstories.org/2015/03/01/the-right-to-energy/.
[194] WRC, 'Safe Access to Fuel and Energy (SAFE)', https://womensrefugeecommission.org/programs/fuel-and-firewood.
[195] UNHCR, 'UNHCR launches global safe energy strategy to benefit millions', 13 May 2014, http://www.unhcr.org/537250486.html.
[196] SAFE, 'United Nations High Commissioner for Refugees (UNHCR)', http://www.safefuelandenergy.org/about/partners.cfm?org=UNHRC.
[197] UNHCR, 'UNHCR Environmental Guidelines', August 2005, http://www.unhcr.org/3b03b2a04.html.
[198] UNHCR, 'Energy Lab', http://innovation.unhcr.org/labs/energy-lab/.
[199] UNHCR, 'UNHCR, The Environment and Climate Change', October 2015, http://www.unhcr.org/540854f49.pdf.
[200] UNHCR, 'UNHCR project brings light, security and fuel-efficient cooking to refugees', 26 January 2012, http://www.unhcr.org/4f21850e6.html.
[201] UNHCR, 'FRAME Toolkit: Framework for Assessing, Monitoring and Evaluating the environment in refugee-related operations', 28 August 2009, http://www.unhcr.org/4a97d1039.html.
[202] UNHCR, 'UNHCR launches global safe energy strategy to benefit millions', 13 May 2014, http://www.unhcr.org/537250486.html.

# Appendix D: Selected Country Policy Contexts

| Policy | Kenya | Tanzania | Ethiopia | Uganda | Sudan |
|---|---|---|---|---|---|
| **Energy status** | | | | | |
| Energy mix | 69% biomass, 22% petroleum products, 9% electricity (2014) | 85.5% biomass, 8.5% petroleum products, 6% electricity (2011) | 88% biomass (2012) | 89.6% biomass, 9.1% petroleum products, 1.3% electricity (2011) | 62% biomass, 6% electricity (2012) |
| Installed capacity and renewable energy share | 2295 MW (2014)<br>63% renewable energy | 1591 MW (2014)<br>35% renewable energy | 2558 MW (2012)<br>91.7% renewable energy | 881 MW (2014)<br>79% renewable energy | 3136 MW (2013)<br>50.8% renewable energy |
| Energy imports/ exports | Exports to Tanzania, imports from Uganda | Imports from Uganda, Zambia and Kenya | Exports to Djibouti and Sudan | Exports to Tanzania and Kenya | Imports from Ethiopia. |
| Electricity access | 32% (2014) | 25% (2014) | 49% (2013) | 10–15% (2012) | 34.5% (2013) |
| **Energy governance** | | | | | |
| Private sector electricity generation | 24% of generation from independent power producers (IPPs) | 41% of generation from private sector | No grid-connected IPPs, some off-grid operators | Majority of power generated by IPPs | No private-sector involvement |
| Generation feed-in tariff | Yes | Yes | No – in development | Yes | No – in development |
| Renewable energy tax incentives | Yes | Yes | Yes | Yes | Yes |
| Renewable energy subsidies/renewable energy funds | No | Yes | No | Yes | No |
| Energy efficiency laws/strategies | Yes – Energy Regulation 2012 | Yes – National Energy Policy Strategy | To be established – Energy Proclamation 2013 | Yes – Energy Efficiency Program 2007 | Yes – National Energy Efficiency Action Plan |
| Targets | Increase renewable energy generation by 5000 MW between 2013 and 2016<br>Electricity access: 70% by 2017 and 100% by 2020 | 2780 MW capacity by 2015, 14% renewable energy generation<br>Electricity access: 30% by 2016 and 75% by 2033 | Increase renewable energy generation to 8–10 MW from 2012 to 2015<br>Electricity access: 75% by 2015 | Increase renewable energy generation to 61% of energy consumption in 2017<br>Electricity access: 22% by 2022 and 100% by 2040 | 12 GW capacity by 2031, including an additional 1582 MW renewable energy generation<br>Electricity access: 45% by 2016 and 80% by 2031 |
| Deforestation policy/ concerns | National Forest Policy 2014 | National strategy and action plan under development to reduce emissions from deforestation and forest degradation | Participatory forest management schemes implemented | Uganda Forest Policy | National Forestry Policy Statement 2006 |
| **Refugee policy** | | | | | |
| Main government body | Department of Refugee Affairs | National Eligibility Committee | Administration for Refugees and Returnee Affairs | The Office of the Prime Minister Department of Refugees | Commissioner for Refugees |
| National refugee policy | Yes | Yes | Yes | Yes | Yes |
| Right to work | Same restrictions as foreign nationals | Right to work (limited activities) in camps once a permit is obtained | Same as foreign nationals; work visa required if not qualified citizens | No restrictions | Work permits required, difficult to obtain, and limited rights |
| Freedom of movement | Restriction on movement | Two-week travel permit required to leave camps | Travel permit required to leave camps | No restrictions on movement | Travel permit required to leave camps |

| Policy | Somalia | Chad | Niger | Nepal | Thailand |
|---|---|---|---|---|---|
| **Energy status** | | | | | |
| Energy mix | 96% biomass, 4% petroleum products (2009) | 90% biomass, 6.5% petroleum products, 3.5% thermal (2013) | 80% biomass, 17% petroleum products, 3% electricity (2012) | 85% biomass, 13% petroleum products, 2% electricity (2012) | 11.4% biomass, 62.3% fossil fuels, 19.4% electricity, 6.9% other renewable energy (2013) |
| Installed capacity and renewable energy share | 80 MW (2012)<br>0% renewable energy | 32 MW (2012)<br>0% renewable energy | 230 MW (2012)<br>0% renewable energy | 763 MW (2012)<br>92.5% renewable energy | 33.3 GW (2013)<br>16.2% renewable energy |
| Energy imports/ exports | None – no centralized grid | None | Imports 83% of electricity from Nigeria | Imports 17% of electricity from India | Imports from Laos and Malaysia |
| Electricity access | 5% (2010) | <4% (2013) | 15% (2012) | 56% (2012) | 99.3% (2009) |
| **Energy governance** | | | | | |
| Private sector electricity generation | Decentralized IPP generation dominates | No private-sector involvement | One state-owned IPP operating | 25% of generation from IPPs (2011) | 47% of generation from independent power producers (IPPs) and small power producers (SPPs) (2012) |
| Generation feed-in-tariff | No | No | No | Yes | Yes |
| Renewable energy tax incentives | No (except for Somaliland) | Yes | No – but in development | Yes | Yes |
| Renewable energy subsidies/ renewable energy funds | No | No – but in development | No – but in development | Yes | Yes |
| Energy efficiency laws/strategies | No (in progress in Somaliland) | No | Yes – 2010 Energy Policy includes energy efficiency strategy | Yes – 2014 Nepal Energy Efficiency Program | Yes – Energy Conservation Act No. 2 (2007) |
| Targets | None | Electricity access: 5% by 2015 | Increase renewable energy share in energy mix to 10% by 2020<br>Electricity access: 3% in rural and 46% in urban areas by 2012 (target missed) | Four GW by 2027, increasing renewable energy share in energy mix to 10% by 2031<br>Electricity access: by 2027, 75% on grid and 20% on hydropower mini-grids | Increase renewable energy share in energy mix to 25% in 2021<br>Reducing electricity imports by 3% in 2016 |
| Deforestation policy/concerns | Heavy deforestation, no national policy | National Forest Policy 2000, National Forest Law 2008 | Heavy deforestation, no national policy | National Forestry Act | National Forest Policy 1985 |
| **Refugee policy** | | | | | |
| Main government body | The National Commission for Refugees and IDPs | Commission Nationale d'Accueil, Réinsertion des Réfugiés et des Rapatriés | Commission Nationale d'Eligibilité au Statut de Réfugiés | Ministry of Home Affairs | Ministry of Interior |
| National refugee policy | In process | No | Yes | No | No |
| Right to work | No policy as yet | No policy as yet | Only if they receive sponsorship, usually from Commission Nationale d'Eligibilité au Statut de Réfugiés | No right to work, can apply as a foreign national but not as a refugee | No right to work, as considered illegal immigrants |
| Freedom of movement | No policy as yet | Travel permit required as per recent decree | No restrictions, same as citizens | Travel permit required to leave camps | Cannot leave camps legally |

| Policy | Afghanistan | Pakistan |
|---|---|---|
| Energy status | | |
| Energy mix | 19% hydropower, 2% diesel, 0.25% thermal, 78% from imports (2014) | 43.8% gas, 39% oil, 11% hydropower, 5.2% coal, 1% nuclear |
| Installed capacity and renewable energy share | 1028.5 MW (2009) | 22,797 MW (2014) |
| Energy imports/exports | Imports from Uzbekistan, Tajikistan, Turkmenistan, Iran | Imports from United Arab Emirates, Kuwait, Oman, France |
| Electricity access | Less than 10% | 93% |
| Energy governance | | |
| Private-sector electricity generation | | |
| Generation feed-in tariff | | Yes |
| Renewable energy tax incentives | | Yes |
| Renewable energy subsidies/ renewable energy funds | | Yes |
| Energy efficiency laws/strategies | | Yes – Pakistan Energy Efficiency and Conservation Act |
| Targets | Reach 65% of households in urban areas | 10% renewable energy or 2700 MW in energy mix by 2015 |
| | Reach 25% of households in rural areas | |
| | Cover at least 75% of total operating costs through user fees by end-2010 | |
| Deforestation policy/concern | Afghanistan faces continued deforestation for illegal timber trade in Pakistan and for domestic fuel purposes; public awareness campaigns have been launched | High deforestation rate but no policy |
| Refugee policy | | |
| Main government body | Ministry of Refugees and Repatriation (MORR) | The States and Frontier Regions (SAFRON) |
| National refugee policy | No, draft law waiting for inclusion in 2015 legislation agenda | Draft law |
| | | National Policy for Afghan Refugees |
| Right to work | | No formal permission but allowed to engage in work in informal economy |

Sources: National energy, deforestation and refugee policies of respective countries.